Was Pythagoras Chinese?
An Examination of Right Triangle Theory in Ancient China

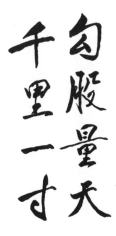

"To use the base and height of a right triangle to measure the heavens, one inch equals a th ͏   ͏
miles."   *Chou pei*

T0204849

*The Pennsylvania State
University Studies No. 40*

Was Pythagoras Chinese?
An Examination
of Right Triangle Theory
in Ancient China

by Frank J. Swetz
and T. I. Kao

The Pennsylvania State University Press
University Park and London

National Council of Teachers of Mathematics
Reston, Virginia

Library of Congress Cataloging in Publication Data

Swetz, Frank.
  Was Pythagoras Chinese?

  (The Pennsylvania State University studies; no. 40)
  Includes translation of the 9th chapter of Chiu chang
suan shu: Kou Ku.
  Bibliography included.
  1. Mathematics, Chinese. 2. Geometry—Early works
to 1800. I. Kao, T. I., joint author. II. Chiu chang
suan shu. Chüan 9. Kou ku. English. 1977. III. Title.
VI. Series: Pennsylvania. State University.
The Pennsylvania State University studies; no. 40.
QA27.C5S95   516'.22'0951   76–41806
ISBN 0–271–01238–2

Contents

Preface

The title of this monograph, while intended to intrigue and attract the otherwise unresponsive reader, was not whimsically conceived. Of course, the historical figure of mathematical fame known as Pythagoras and born on the island of Samos in the sixth century B.C. was Greek and not Chinese. But there is another "Pythagoras" equally famous. He is the man who first proved the proposition that "the sum of the squares of the legs of a right triangle is equal to the square of the hypotenuse." For hundreds of years this theorem has borne the name of Pythagoras of Samos, but was he really the first person to demonstrate the universal validity of this theorem? The issue is controversial. Seldom are mathematical discoveries the product of a single individual's genius. Often centuries and thousands of miles separate the appearance and the isolated reappearance of the same mathematical or scientific theory. It is now acknowledged that the "Pascal Triangle" method of determining the coefficients of a binomial expansion was known in Sung China three hundred years before Pascal was born and that the root extraction algorithm credited to the nineteenth-century British mathematician W.G. Horner was employed by Han mathematicians of the third century A.D. If, then, these mathematical processes are to bear the names of the men or women who devised them, surely "Pascal" and "Horner" were Chinese. So too might such an argument be posed for the Pythagorean Theorem on the basis of evidence contained in ancient Chinese mathematics texts. It is the main purpose of this monograph to expose and examine this evidence.

With the publication of the third volume of *Science and Civilization in China* by Joseph Needham (31),* a new era of interest concerning ancient China's contributions to the field of mathematics has been initiated. Not since the late nineteenth and early twentieth centuries have Westerners so consciously sought to explore and understand the significance of historical China's mathematical thinking. Most classical Chinese mathematics texts remain untranslated and unknown in the West. Mathematical historians have often ignored or depreciated China's contributions to mathematical knowledge. Fortunately, this situation seems to be changing, as demonstrated by the recent appearance of Ulrich Libbrecht's analysis of the *Shu-shu chiu-chang* (25).

* Numbers in parentheses refer to Reference entries.

The authors of this monograph wish to contribute to this change by investigating ancient Chinese theory and application concerning the right triangle and by submitting the following annotated translation of the *Kou-ku* of the *Chiu chang suan shu,* the richest source of problems from antiquity dealing with the right triangle. This work was not undertaken solely as an academic pursuit but rather to appease an aroused curiosity concerning sino-mathematical achievements, a curiosity whetted by the writings of such scholars as Needham and Mikami (29). Our translation efforts were confined to the last chapter of the *Chiu chang* primarily with the desire to reveal and better understand the geometric-algebraic solution methods employed by the ancient authors in their work with right triangles and, secondly, in light of our limitations, as a matter of expedience. The *Chiu chang suan shu,* most influential of all Chinese mathematics texts, remains, itself, untranslated into English, although translations do exist in Russian (1) and German (41). To the author's knowledge, this text provides the first complete English translation of the ninth chapter of the *Chiu chang.*

The particular copy of the text employed in the translation is a reprint of the *Chiu chang* as contained in the *Yung lo ta tien,* a Ming encyclopedia. The questions and their solution methods were checked against those given in *Hsiang chieh chiu chang suan fa* (A Detailed Analysis of the Mathematical Methods in the "Nine Chapters") by Yang Hui (1261a) (45).

It is hoped that our efforts may be of some service to both the casual and serious students of the history of mathematics and scientific thought. Teachers of mathematics will find the concrete-based problem-solving methodology of the ancient Chinese pedagogically appealing for classroom use, particularly in mathematics laboratory situations. The historical insights provided certainly can be used to enrich lecture presentations. In the final offering, this work does not pretend to resolve the initial question it poses but rather to amplify it. While providing information on ancient Chinese mathematical accomplishments, our findings will raise more questions than they answer.

The authors wish to thank Professor Shih-chuan Chen of Penn State's Capitol Campus for his personal translation and calligraphic rendering of the *Chou pei* quotation, Robert Paradine of The Pennsylvania State University Press for his encouragement and assistance in the production of this work, and Mrs. Margaret Alexander of the Mathematics Department at Capitol Campus for her expert typing of the final manuscript.

Harrisburg, Pennsylvania   F.J. Swetz
                                                                                                      T.I. Kao

Monetary Units and Measures

Monetary
    tael = 37.8 grams of silver
    wen = small coin
    kuan = 1000 wen

Weight
    picul = 60.48 kilograms

Capacity
    tou = 10.3 liters
    hu = 10 tou

Land Measures
    li (mile) = 300 pu
    pu (pace) = 5 ch'ih
    chang = 10 ch'ih
    ch'ih (foot) = 10 ts'un
    ts'un (inch) = 3.58 centimeters
    mou (acre) = 240 square pu

1

Perspectives

The Pythagorean Theorem: Historical Significance

If one asked a knowledgeable audience "What, in the early history of the human race, were the milestones of mathematical accomplishment?," a variety of responses would be forthcoming. Some would agree on the importance of certain events, others would offer contention. The debate would become more involved if this audience were then requested to select the most profound accomplishment on the basis of its level of intellectual achievement and its eventual consequences for mankind. The list might then narrow down to two or three decisive events and most likely among these would be the derivation of the "Pythagorean" Theorem, for the appearance of this theorem marks the first known intellectual jump between the confines of empirical speculation and the limitless bounds of deductive reasoning.

From the earliest times, man has perceived his spatial environment in terms of vertical and horizontal—a tree grows vertically to the horizontal plane of the earth, a companion stands vertical to the surface that supports him. These vertical-horizontal relations were manifestations of the action of gravity on terrestrial objects. No doubt, man rapidly learned to use this relationship to his advantage; thus for maximum efficiency and security, the supporting pole for a shelter should be placed vertically to the ground. While nature supplied the primal example of perpendicularity, man soon must have realized that elements of the vertical realm always met with elements of the horizontal realm in the same manner or visual pattern, forming what we know as a right angle. Once this concept was grasped, the potential of this union could be utilized in a wider variety of human endeavors such as the stringing of an arrow to a bow or erecting permanent structures of wood and stone. Perpendicularity could now be translated from the fixed vertical-horizontal constraints of nature and used to human benefit. Some of the first "scientific instruments" devised incorporated in their functioning the use of right angles. With an understanding of perpendicularity, a pole fixed in the ground could become a gnomon, an instrument for determining the solstices of the sun and thus the changing of the seasons of the year or a sighting marker to lay out a straight wall. The pyramids of Egypt stand in mute testimony to the surveying applications of the right angle. Both the construction of the Egyptian plumb bob level and the rope loop square were based on an understanding of the right angle and right triangle.[1]

Eventually, an empirically based formulation of the relationship of perpendicularity emerged in the form of a rule that related the three sides of a right triangle—the sum of the square of the legs is equal to the square of the hypotenuse. With this statement, the very essence of the right triangle and perpendicularity is exposed. Two millennia before the modern era, Babylonian tables of number triples, some as impressive as $3367:3456:4825$, indicated that the society of the time knew of the right triangle rule (34). But tables are cumbersome and limited in use, especially when they are made out of stone. No matter how proficient the Babylonians were in their systems of measurement and methods of calculation concerning right triangles, their tabulations were only for certain specific right triangles. As long as the rule remained unproven in universal applications, there always would remain the possibility that a right triangle could be encountered whose sides did not satisfy the accepted relationship. With the validity of the right triangle rule in doubt, the development of further rules that might use the right triangle rule as a first principle or premise would also be questionable. History books tell us that some time in the sixth century B.C. Pythagoras of Samos derived a proof for this theorem. The appearance of this theorem then permitted the concept of perpendicularity to readily be accepted in the formulation of further scientific and mathematical rules until the present day when almost no branch of mathematics is free from the concept of perpendicularity or its n-dimensional analog, orthogonality.[2] Thus the conception of the "Pythagorean Theorem" is one of the most profound accomplishments in the history of mathematics.

Early Evidence of Chinese Work with the Right Triangle

The noted humanistic historian of science Jacob Bronowski supports this claim:

> To this day, the theorem of Pythagoras remains the most important single theorem in the whole of mathematics. That seems a bold and extraordinary thing to say, yet it is not extravagant; because what Pythagoras established is a fundamental characteristic of the space in which we move, and it is the first time that it is translated into numbers. And the exact fit of the numbers describes the exact laws that bind the universe. In fact, the numbers that compose right-angled triangles

have been proposed as messages which we might send out to planets in other star systems as a test for the existence of rational life there.[3]

Bronowski in his consideration of the "Pythagorean Theorem" cites a simple but elegant proof for this theorem, which he speculates Pythagoras might have employed.[4] In this proof a square is formed by four congruent right triangles, the side of the square being the hypotenuse of the triangle; the four triangles are then rearranged to comprise two smaller squares, the measures of whose respective sides are equal to the legs of the right triangle, thus demonstrating rather concretely that the sum of the squares of the legs of a right triangle is equal to the square of the hypotenuse (see Fig. 1.1). The proof is aesthetically appealing and Bronowski's case for Pythagorean authorship is convincing; however, no record of such a proof comes to

Figure 1.1

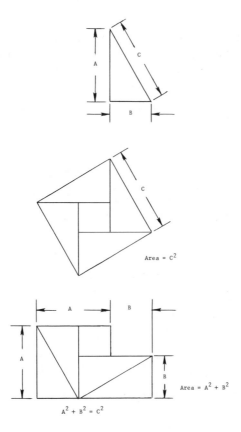

us from Greek sources. Further, the historian of ancient Greek mathematics Thomas Heath states that this type of proof was alien to Greek geometrical methods and modes of thinking.[5] While Greek origins for such a proof remain in doubt, it is known to have appeared in the works of the Indian mathematician Bhāskara in 1050. It seems from available evidence that Bhāskara was influenced by earlier Chinese mathematical works, and if we search back farther into the history of Eastern mathematics, this proof appears in an ancient Chinese mathematical text, the *Chou pei suan ching*[a]* (*The Arithmetical Classic of the Gnomon and the Circular Paths of Heaven*). This book is the oldest Chinese mathematics text known. While the exact date of its origin is controversial, with estimates ranging as far back as 1100 B.C., it can generally be accepted on the basis of astronomical evidence that much of the material in the book was from the time of Confucius, the sixth century B.C., and its contents would reflect the mathematical knowledge accumulated in China until that time.

The diagram that describes the proof we are discussing was so well known in China that it had a special name, *hsuan-thu*[b] (see Fig. 1.2). The following passages accompany the *hsuan-thu* in the *Chou pei* and reveal Chinese knowledge of and concern with the right triangle at this early date:

1. Of old, Chou Kung addressed Shang Kao, saying, "I have heard that the Grand Prefect (Shang Kao) is versed in the art of numbering. May I venture to inquire how Fu-Hsi anciently established the degrees of the celestial sphere? There are no steps by which one may ascent the heavens, and the earth is not measurable with a footrule. I should like to ask you what was the origin of these numbers?"

2. Shang Kao replied, "The art of numbering proceeds from the circle and the square. The circle is derived from the square and the square from the rectangle (lit. T-square or carpenter's square).

3. The rectangle originates from the fact that $9 \times 9 = 81$ (i.e., the multiplication table or properties of numbers as such).

4. Thus, let us cut a rectangle (diagonally), and make the width 3 (units) wide, and the length 4 (units) long. The diagonal between the two corners will then be 5 (units) long. Now after drawing a square on this diagonal, circumscribe it by half-rectangles like that which has been left outside, so as to form a (square) plate. Thus the (four) outer half-rectangles of width 3, length 4, and diagonal 5, together make two rectangles (of area 24); then (when this is subtracted from the square plate of area 49) the remainder is of area 25. This (process) is called 'piling up the rectangles' (*chi chu*).

5. The methods used by Yu the Great in governing the world were derived from these numbers."

6. Chou Kung exclaimed, "Great indeed is the art of numbering. I would like to ask about the Tao of the use of the right-angled triangle" (lit. T-square).

---

* Superscript letters refer to the equivalent Chinese characters to be found in the glossary at the end of this work.

Figure 1.2

7. Shang Kao replied, "The plane right-angled triangle (laid on the ground) serves to lay out (works) straight and square (by the aid of) cords. The recumbent right-angled triangle serves to observe heights. The reverse right-angled triangle serves to fathom depths. The flat right-angled triangle is used for ascertaining distances.
8. By the revolution of a right-angled triangle (compasses) a circle may be formed. By uniting right-angled triangles squares (and oblongs) are formed.
9. The square pertains to earth, the circle belongs to heaven, heaven being round and the earth square.[6] The numbers of the square being the standard, the (dimensions of the) circle are (deduced) from those of the square.
10. Heaven is like a conical sun-hat. Heaven's colors are blue and black, earth's colors are yellow and red. A circular plate is employed to represent heaven, formed according to the celestial numbers; above, like an outer garment, it is blue and black, beneath, like an inner one, it is red and yellow. Thus is represented the figure of heaven and earth.
11. He who understands the earth is a wise man, and he who understands the heavens is a sage. Knowledge is derived from the straight line. The straight line is derived from the right angle. And the combination of the right angle with numbers is what guides and rules the ten thousand things."
12. Chou Kung exclaimed, "excellent indeed."[7]

From this dialogue it is evident that the geometrical priorities of ancient China evolved around the land mensuration needs of an agrarian society. The reference to Yu the Great pertains to a lengendary figure who was considered the father of Chinese mathematics and the "patron saint" of hydraulic engineers. The Chinese society of this time was feudalistic in its social organization.[8] Its survival was contingent upon functioning systems of irrigation canals, retaining dikes, and protective fortifications. Thus in early China geometric knowledge focused on the principles of basic construction and mensuration, and in particular appropriate uses of the right triangle in surveying situations. Geometric theory was pragmatically inclined and empirically based.

Although the *Chou pei* contains discourses on the properties and uses of the right triangle, they are obscured by incorporation into a mystical cosmology. In contrast, later mathematical works are conspicuously free of mystical connotations, but they are also equally devoid of supporting theoretical justifications. For the humanistic scholars of the Middle Kingdom, mathematics was a *shu*,[c] a technique, necessary for the ends it could accomplish and unworthy, in itself, as an object of speculation; in fact, mathematicians enjoyed a social status equivalent to a common clerk. The oldest extant mathematical work that provides valuable insights into the geometric computational ability of early China involving situations dealing with the properties of right triangles is the *Kou-ku*[d] chapter of the *Chiu chang suan shu*[e] (*Nine Chapters on the Mathematical Art*). Although our main interest centers on the last chapter of the *Chiu chang*, the complete work is unique in the history of mathematics, deserving at least a cursory examination.

The *Chiu chang suan shu:* Its Content and Form

Dating of ancient Chinese literature is extremely difficult. Despotic emperor Shih Huang-ti[f] of the Ch'in dynasty (221–207 B.C.) originated the first burning of books in 213 B.C. Scholars of the following Han period (206 B.C.–A.D. 220) were forced to transcribe China's literary and scientific traditions from memory or remaining scroll fragments. Hyperbole was often used in attributing a work to an ancient sage with the design of increasing its prestige; forgery was a common practice. As a result of such phenomena, the dating of the *Chiu chang* has at times been confused. However, it is accepted today to be a product of the late Ch'in and early Han dynasties (third century B.C.) and was probably first written by one Chang Tshang,[g] who made use of earlier works then in existence. The version that survives to the present day is a commentary prepared by a certain Liu Hui[h] in approximately A.D. 250. Thus the content of this text is a summary of the mathematical knowledge possessed in China up to the middle of the third century, its nucleus having been assembled two hundred years before the modern era. Through the ages, the text has been scripturally preserved in the Confucian tradition with each scribe-mathematician copying the previous edition and adding his commentary to it (see Fig. 1.3). The first westerner to translate parts of the text into English was Alexander Wylie, a scholar and competent mathematician in the employ of the Manchu government (43,44). Wylie marveled at his findings.

The *Chiu chang* consists of nine distinct sections or chapters: three involving surveying and engineering formulas, three devoted to problems of taxation and bureaucratic administration, and the remaining three to specific computational techniques. These textual divisions testify to the bureaucratic concerns of the time. Instruction is provided through the presentation of specific problems and rules for obtaining their solutions. Whereas the early authors did not possess a system of algebraic notation, all exposition is in a literary form often confusing to the modern reader. Originally, little indication was provided as to just how the solution rules were derived, but the general context is algebraic-arithmetic suggesting an empirical methodology based on a concrete manipulation of models. Later commentaries attempt to rectify this omission by providing solution explanations. A brief summary of the nine chapters follows:

I. *Fang thien*[i] (land surveying).[9] Various formulas are given for elementary engineering work. They include correct computational rules for finding the area of rectangles, triangles, trapezoids, and circles with π being approximated by 3. A trapezoidal approximation for the area of a segment of a circle is given by $(S/2)(C + S)$. C being the measure of the length of the chord involved and S the measure of the length of the sagitta of the segment. (A discussion on the adequacy of such an approximation could serve as an interesting present-day classroom exercise for teachers of mathematics.)

This section also reveals an arithmetic of fractions employing the principles of lowest common denominators and greatest common factors in accord with modern practices. Such an approach to operations with fractions was not utilized in Europe until the fifteenth and sixteenth centuries. The utilitarian nature of the *Chiu chang*'s presentation fosters an elaborate treatment of common fractions at the expense of decimal fractions, which are also used in the work, but to a lesser extent.

Approximation of π by 3 is common in ancient mathematical texts. Wylie, while in the process of translating this work, queried his Chinese colleagues on the adequacy of using this value of π. He was politely informed that the scholars of the ancient period of authorship intended to supply a sufficient approximation rather than a more accurate but unwieldy one. When we remember the nature of this work and the fact that similar Egyptian and Babylonian writings chose 3 over more exact values, credibility can be given to this theory. Needham's research would also seem to bear this out.[10]

II. *Su mi*[j] (millet and rice). This chapter treats questions of simple percentage and proportions. It was originally used for grain exchange and included tables of conversion for weights of varied grains in terms of rice but later expanded to include a variety of problems whose solutions depended on a use of proportions.

III. *Tshui fen*[k] (distributions by progressions). Problems involving taxation, partnership, and arithmetical and geometrical progressions are presented. Their solutions are arrived at by methods of proportion and the "rule of three."

Although the rule of three has generally been considered of Indian origin,[11] the *Chiu chang*'s exposition of this technique predates Sanskrit versions (ca. 628). This rule, held in high regard in the mercantile society of medieval Europe, is simply a proportion involving three knowns from which an unknown quantity is found. A Chinese example involving this principle goes as follows:

Two and one-half picules of rice are purchased for ³⁄₇ of a tael of silver. How many can be purchased for 9 taels?

Employing modern symbolism and the rule of three, the solution can easily be found:

$$\frac{2\frac{1}{2}}{\frac{3}{7}} = \frac{x}{9}$$

$$\frac{35}{6} = \frac{x}{9}$$

$$x = \frac{(35)(9)}{6}$$

$$x = 52\frac{1}{2} \text{ piculs}$$

Typical of the problems concerning series found in this section is one involving a farmer's dilemma:

> A cow, a horse and a goat were in a wheat field and consumed some stalks of wheat. Damage of 5 tou of grain were asked by the wheat field's owner. If the goat ate one half the number of stalks as the horse, and the horse ate one half of what was eaten by the cow, how much should be paid by the owners of the goat, horse and cow respectively?

A modern student with a minimum of algebraic knowledge would interpret the problem symbolically prior to finding a solution for the expression:

$$\tfrac{1}{4}x + \tfrac{1}{2}x + x = 5$$

IV. *Shao kuang*[1] (diminishing breadths). This section contains twenty-four problems on land mensuration. Some of the examples given illustrate the Chinese process for finding square and cube roots. Wylie noted that the rules prescribed for solution were very similar to those given in contemporary (1876) English school books.

Square and cube roots of numbers were extracted through a computational procedure known as the "Celestial Element Method." The concept behind this procedure paralleled the method devised by W.G. Horner in the nineteenth century. Twenty centuries earlier, the Chinese in the manipulation of their counting rods were employing Horner's method in solving equations such as $x^3 = 1,860,867$ and $x^2 + 34x = 71,000$. Although later mathematical historians were to affirm Wylie's original observations, it was not until 1955, when Wang and Needham's published research (42) revealed and analyzed this method in detail, that the fact was accepted—these two root extracting processes were the same in principle.

The Celestial Element Method—so named because the Chinese character that represented the word heaven, *t'ien*,[m] also was used to represent the unknown—was derived from the geometrically established relation $(a + b)^2 = a^2 + 2ab + b^2$. Theon of Alexandria (ca. 390) used a similar procedure for arriving at his rules for finding square roots.[12] Although the *Chiu chang* apparently deals with integral roots in its solutions, there is evidence that scholars of the time could extract roots to several decimal places. Problems 9 and 10 of this section as solved in a first-century commentary reveal that the Chinese sometimes extracted square and cube roots employing the computational technique whereby $\sqrt[n]{m}$ could be replaced by $\sqrt[n]{mp/p}$. If p is allowed to be a power of ten so that

$$\sqrt[n]{m} = \frac{\sqrt[n]{m\,10^{kn}}}{10^k}$$

the method assists in the decimalization of roots and permits a greater accuracy in root extraction. Han mathematicians in obtaining solutions for problems such as $\sqrt{0.35}$ would transform them into decimal fraction equivalents such as $\sqrt{350/1000}$ hence $\sqrt{350000}/1000$ and thus could readily extract the square root of 350,000, using their counting rods and the Celestial Element Method. A generalized version of this technique, transmitted westward by Hindu and Moslem agents, was used in Europe in the late middle ages in the computation of square root tables.

Some controversial problems found in this section concern work with unit fractions. These problems confused early researchers into conjecturing that the ancient Chinese employed a system of unit fractions similar to that of the Egyptians.

> There is a (rectangular) field whose breadth is one pu and a half and a third. If the (area) is one mou, what is the length of the field?
>
> Answer: $130^{10}/_{11}$ pu.

The solution rule indicates the Chinese facility for working with fractions:

> As there is 3 in the denominator, we have to change 1 to 6, a half to 3 and one-third to 2. Adding these together we get 11, which will be made the fa or divisor. Set down the 240 square pu (1 mou) of the field, where every 1 is to be counted as 6, and so this number 6 is to be multiplied. The product is the shih or dividend. Thus, carrying out the division we get the number of pu contained in the length.

V. *Shang kung*[n] (consultations of engineering works). Here we are provided with formulas for the computation of volumes encountered in constructing dikes or fortifications; for prisms, cylinders, tetrahedrons, pyramids, circular cones, and various truncated solids. One of the more unusual formulas given is that for the volume of a truncated triangular right prism, $(b_1 + b_2 + b_3)DL/6$, where b_1, b_2, b_3 are measures of the three breadths of the wedge, D is the measure of the depth, and L the measure of the length. Legendre is usually credited with having devised this formula in his *Eléments de Géométrie* published in 1794.[13] Also noteworthy is the formula $\frac{1}{6}$ *abh* for the volume of a tetrahedron whose two opposite edges, *a* and *b*, are at right angles, and where *h* is the measure of the height of the common perpendicular. It appears highly likely that formulas such as these were the end product of experimentation employing concrete models of the structures under concern.

VI. *Chun shu*[o] (impartial taxation). Consideration is given to the distribution of taxes and the various difficulties in transporting the taxes, as paid for with grain, to the capital and pursuit and interception problems. Typical of the taxation problems is the following:

> Four counties are required to furnish wagons to transport 250,000 hu of grain to a depot. There are 10,000 families in the first county, which is 8 days' travel to the station; 9,500 families in the second county at a distance of 10 days' travel; 12,350 families in the third county 13 days away, and 12,200 families in the last county which is 20 days' travel from the depot. The total number of required wagons is 1,000. How many wagons are to be provided by each county according to the size of its population and the distance from the depot? (The levy of wagons is directly proportional to the population of the county and inversely proportional to the distance from the depot.)

Pursuit problems, the nemesis of many a school boy, were rather late in arousing European interest. Smith lists as one of the first Western sources of such problems Alcuin of York's *Propositiones ad Acundas Juvenes* (ca. 775), in which a hound pursuing a hare problem is given.[14] The *Chiu chang's* version of this problem is:

> A hare runs 50 pu ahead of a dog. The latter pursues the former for 125 pu. When the two are 30 pu apart, in how many pu will the dog overtake the hare?

VII. *Ying pu tsu*[p] (excess and deficiency).[15] This section is devoted to the Chinese algebraic technique commonly called in the West the "rule of false position," used to solve equations of the form $ax - b = 0$. Although the solution of such problems now would be considered trivial, for centuries before the adoption of a manipulative symbolism, their literal presentation posed a high degree of difficulty. Let us examine the Chinese method in some detail armed with modern algebraic notation.

We wish to solve $ax - b = 0$; let us make two guesses as to the value of x, g_1, g_2. These guesses result in two failures f_1, f_2, that is:

(1) $$ag_1 - b = f_1$$

(2) $$ag_2 - b = f_2$$

Subtracting the second expression from the first, we have:

(3) $$a(g_1 - g_2) = f_1 - f_2 \quad \text{or} \quad a = \frac{f_1 - f_2}{g_1 - g_2}$$

Multiplying eq. 1 by g_2 and eq. 2 by g_1, we obtain

(4) $$ag_1g_2 - bg_2 = f_1g_2$$

(5) $$ag_1g_2 - bg_1 = f_2g_1$$

Subtracting eq. 5 from eq. 4 and simplifying, we arrive at:

$$b(g_1 - g_2) = f_1g_2 - f_2g_1$$

Combining this expression with the one given for a in eq. 3, we have

$$x = \frac{b}{a} = \frac{f_1g_2 - f_2g_1}{f_1 - f_2}$$

Thus we have solved for x in terms of our "guesses" and "failures." The Chinese guesses were chosen such that one resulted in too large an answer and the other in too small a value for the unknown; therefore, the method was known as (excess and deficiency). The following problem from this section is introductory in nature and the application of the rule is obvious:

> A certain number of people are buying some chickens. If each person pays 9 wen there is a surplus of 11 wen, and if each person pays 6 wen there is a deficiency of 16 wen. Find the number of people and the cost of the chickens.

> Answer: 9 people and the chickens cost 70 wen.

Solutions for problems given later in this section (e.g., problems 9–20) require a judicious use of guessing based on a knowledge of the subjects in question.

From the thirteenth century, a variation of this method was widely employed in Europe under the name *Regula Falsae Positionis*. It was formerly believed that this rule was transmitted to Europe from India by Arab mathematicians; careful scrutiny of Sanskrit texts reveals the earliest indication of such a method in the work of *Brahmagupta* (ca. 628). The Egyptian *Rhind Papyrus* reveals another method of false position from antiquity; however, this differs from the Chinese version and its European adaption.

VIII. *Fang ch'eng*[q] (the way of calculating by tabulation). Eighteen problems are presented dealing with the solution of systems of simultaneous linear equations.

Counting rod techniques require their user to establish a matrix representing the numerical coefficients of the given systems of equations.[16] Elementary column and row operations consistent with modern practices were performed on the matrix to obtain a solution. The examination of a problem from this section would prove enlightening:

> There are three grades of corn. Neither two baskets of the first grade, three baskets of the second, nor four baskets of the third grade taken separately comprise a full required measure. If, however, one basket of the second grade were added to the first grade corn, one basket of the third grade corn added to the second, and one basket of the first grade to the third grade baskets, then the grain would comprise one required measure in each case. How many measures of the various grades does each mixed basket contain?

That is,

$$2x + y = 1$$
$$3y + z = 1$$
$$x + + 4z = 1$$

The matrix formed is:

1		2	1st grade (x)
	3	1	2nd grade (y)
4	1		3rd grade (z)
1	1	1	required measure

A "zero" coefficient was indicated by a blank in the computational matrix. This particular problem requires its solver to subtract a number from nothing! A rule is provided for such a possibility—distinguishing the Chinese as the first known society to have developed an algebra of negative numbers, *Chang-fu-shu*[r] (positive and negative). Similarly, coefficients in the matrix could also be negative in themselves. They were indicated by red calculating rods in this scheme. Black rods represented positive numbers. Previously the earliest accepted use of negative numbers in a society had been credited to Indian mathematicians (ca. 630).[17] The existence of such problems in the *Chiu chang* discredit this claim. While certainly advanced in their computational techniques, the Chinese failed to develop this method further into a theory of determinants. An heir to this early work, the Japanese mathematician Seki Kōwa, finally formalized a theory of determinants in 1683, predating a similar European achievement by ten years.

The solution of a system of four equations in five unknowns is considered in the last problem of the section, indicating a Chinese investigation of indeterminate equations at an early date. Chinese mathematicians of a later period were to refine indeterminate analysis and christen this area of study *Ta yen chiu shu*[s] or "searching for unity." Typifying the concern of Chinese scholars were problems such as that of the "Hundred Fowls" (ca. 475):

If a cock is worth 5 coins, a hen 3 coins and three chickens together only 1 coin, how many cocks, hens and chickens, together totaling 100, may be bought for 100 coins?

IX. *Kou ku* (right angles). Twenty-four problems are presented concerning the properties of right triangles. The accurate surveying of the farming lands of China demanded an understanding and application of the "Pythagorean" Theorem at a very early date. Since Chapter 2 is devoted to a complete translation of this section, no further consideration of it is given here.

The contents of the *Chiu chang* are a testimonial to the high degree of mathematical insight and computational proficiency possessed and utilized in early China. Its 246 problems make it richer in scope than any previously discovered Egyptian or Babylonian texts, and similar Greek problem collections exist only for later periods of history. While the knowledge that is evidenced in the first eight sections would provide a substantial case for acclaiming ancient China's mathematical eminence, that remaining in the ninth section establishes a pre-eminence in the consideration of the properties and the applications of the right triangle.[18]

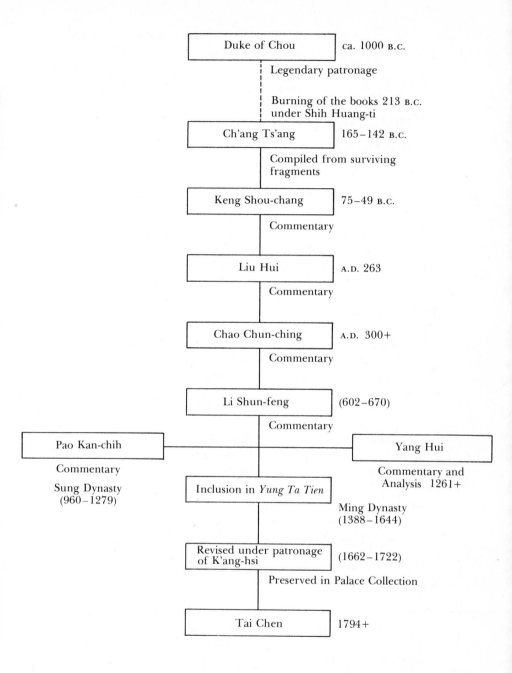

Figure 1.3   *Lineage and chronology of partial influences and events*
affecting the Chiu Chang Suan Shu

2

The *Chiu Chang*'s Problems Involving Right Triangles

The following annotated translation of the ninth chapter of the *Chiu chang* together with its solution commentaries reveal the geometric-algebraic methods of the Han scholars and should foster an appreciation of the problem-solving capabilities of these early mathematicians. Analyses of the problem situations highlight the importance of the "Pythagorean" proposition and its applications, both of which are so often obscured in the modern world by the presence of more glamorous mathematical concepts.

Our translation is intended to provide such an examination of the *Kou-ku* of the *Chiu chang*. The twenty-four problems of the last chapter as ascribed to Liu Hiu have been translated in their entirety. The copy of the text employed in this effort is a reprint (9) of the *Chiu chang*, scripturally preserved by Pao Kan-chu[t] of the late Sung dynasty, contained in the *Yung lo ta tien*,[u] a Ming encyclopedia,[1] and revised under the sponsorship of Ching Emperor K'ang-shi[v] (1662–1722). The primary commentary on Liu's work is supplied by Li Shun-feng[w] of the T'ang dynasty. Illustrations supplementing the text are considered a product of Ching scholars, as the original diagrams of Liu's text were lost centuries ago. They do, however, represent the Han computational processes and are worth noting.

[1.][2] Given: kou = 3 ch'ih, ku = 4 ch'ih. What is the length of hsien?
        Answer: hsien = 5 ch'ih.
[2.] Given: hsien = 5 ch'ih, kou = 3 ch'ih. What is the length of ku?
        Answer: ku = 4 ch'ih.
[3.] Given: ku = 4 ch'ih, hsien 5 ch'ih. What is the length of kou?
        Answer: kou = 3 ch'ih.

kou ku: In a right triangle, the short side adjacent to the right angle is called kou. The longer side adjacent to the right angle is called ku. The side opposite the right angle is called hsien. Kou is shorter than ku, and ku is shorter than hsien. We are going to use this relationship, so we explain it at this point in order to assist the reader in his understanding of the solution methods.

Method (problem 1): Add the square of kou and ku. The square root of the sum is equal to hsien.

Figure 2.1

圖 求 互 弦 股 句

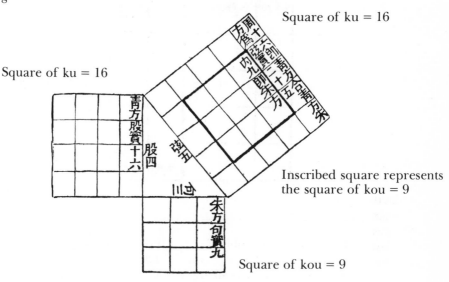

Square of ku = 16

Square of ku = 16

Inscribed square represents the square of kou = 9

Square of kou = 9

Square of ku equals (kou + hsien) × (hsien − kou) = 16

Method (problem 2): The length of kou is equal to the square root of the difference of the squares of hsien and ku.

Figure 2.2

圖 矩 之 實 句

Square of ku = 16

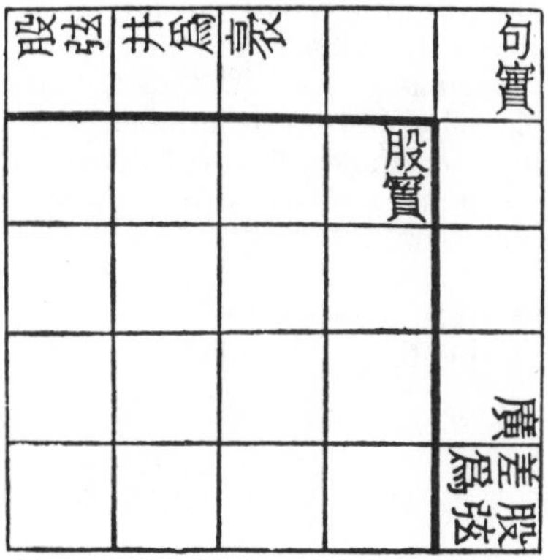

Method (problem 3): The length of ku is equal to the square root of the difference of the squares of hsien and kou.

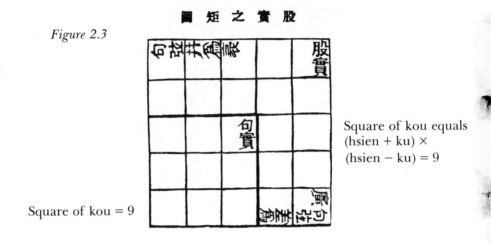

Figure 2.3

圖 矩 之 實 股

Square of kou equals (hsien + ku) × (hsien − ku) = 9

Square of kou = 9

Commentary:* The Chinese characters for kou and ku originally were used to designate "leg" and "thigh." *Hsien*[x] in a nonmathematical context designates a string strung between two points such as a lute string. The 3, 4, 5 right triangle was apparently known in many ancient societies and appreciated for its utilitarian character. The *harpedonaptai* (rope-stretcher), priest-surveyors of the Nile Valley, employed it in their work. Texts from the Seleucid period of Babylonian history (after 310 B.C.) contain problems concerning this particular triangle.[3] The earliest known Chinese consideration of the right triangle, that embodied in the *Chou pei*, also concerns the 3, 4, 5 right triangle:

> The art of numbering proceeds from the circle and the square. The circle is derived from the square and the square from the rectangle.
> The rectangle originates from (the fact that) $9 \times 9 = 81$ (i.e., the multiplication table or properties of numbers as such).
> Thus, let us cut a rectangle (diagonally), and make the width (kou) 3 (units) wide, and the length (ku) 4 (units) long. The diagonal (hsien) between the (two) corners will then be 5 (units) long . . . .[4]

Chao Chu-ching,[y] first known commentator on the *Chou pei*, explains these passages as indicating the necessity of acquiring number-

---

* The authors' comments throughout this chapter are distinguished from the translation by rules.

28

theoretic knowledge before proceeding to geometrical investigations.[5] In light of this sentiment and the frequent recurrence of the 3, 4, 5 triangle in the old works, the conjecture might easily be raised that early number-theoretic investigations of the 3, 4, 5 triple resulted in the discovery of the "Pythagorean" equation.[6] Did the manipulation of numbers by ancient priest-mathematicians seeking mystical properties result in the discovery of the "Pythagorean" proposition? This issue will remain conjecture until further work on the nature of ancient mathematics is realized.

Several remaining illustrations of the *Kou-ku* incorporate 3, 4, 5 right triangles, although the problems in question contain a different measurement ratio. Presumably, the commentators found this triangle a satisfying didactical device to employ in their "derivations" of solution rules.

[4.] Given: A wooden log of diameter 2 ch'ih 5 ts'un, a 7 ts'un (thick) board is to be cut from the log. What is the (maximum) width of the board?
Answer: 2 ch'ih 4 ts'un.
Method: From the square of 2 ch'ih 5 ts'un subtract the square of 7 ts'un. The width of the desired board is equal to the square root of the difference.
Explanation: 2 ch'ih 5 ts'un equal hsien. 7 ts'un equal kou; the width of the board is ku.

Commentary: A direct application of the "Pythagorean" Theorem (see Fig. 2.4).

Figure 2.4

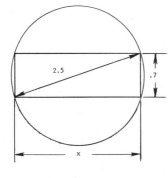

$$x^2 + (.7)^2 = (2.5)^2$$
$$x = \sqrt{(2.5)^2 - (.7)^2}$$
$$x = 2.4 \text{ ch'ih}$$

[5.] Given: A tree of height 20 ch'ih has a circumference of 3 ch'ih. There is an arrow-root vine[7] which winds seven times around the tree and reaches to the top. What is the length of the vine?
Answer: 29 ch'ih.

Method: Ku equals the product of 7 and 3 ch'ih, 21 ch'ih. Kou equals the height of the tree, 20 ch'ih. The length of the vine is equal to the hsien.

Explanation: One can understand the problem by winding a string around a writing brush. When unwound, it can be seen that one unit of the resulting helix forms a triangle. For the required problem, we consider the ku to be the product of 7 and the given circumference and kou is equal to the height of the tree, then the only unknown is hsien, which is the (required) length of the vine.

Commentary: The authors of this work realize that the length of a helix could be determined by unwinding the helix from the surface of the cylinder upon which it is circumscribed. The trace of the helix then becomes the hypotenuse of a right triangle whose legs are the height of the cylinder and the product of the circumference of the cylinder and the number of revolutions circumscribed by the helix. (See Fig. 2.5.)

Figure 2.5

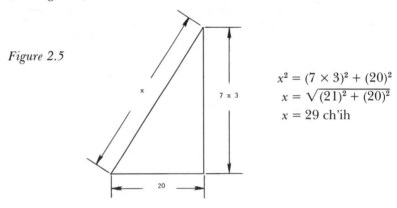

$$x^2 = (7 \times 3)^2 + (20)^2$$
$$x = \sqrt{(21)^2 + (20)^2}$$
$$x = 29 \text{ ch'ih}$$

[6.] Given: In the center of a square pond whose side is 10 ch'ih grows a reed whose top reaches 1 ch'ih above the water level. If we pull the reed toward the bank, its top is even with the water's surface. What is the depth of the pond and the length of the plant? (Fig. 2.6a)

Answer: The depth of the water is 12 ch'ih and the length of the plant is 13 ch'ih.

Method: Find the square of half the pond's width, and from it subtract the square of 1 ch'ih. The depth of the water will be equal to the difference divided by twice the height of the reed above water (1 ch'ih). To find the length of the plant we add 1 ch'ih to the result.

Explanation: The problem can be solved by considering half the width of the pond to be kou, the depth of the water to be ku, and the length of the plant to be hsien. Kou and the difference of ku and hsien are known. Therefore we can find the square of kou or 5 ch'ih and the rectangle (product) of the difference and sums of ku and hsien is known (where) the difference of ku and hsien is 1 ch'ih; (from this product) subtract the square of 1 ch'ih. The remainder will be the product of twice the difference of ku and hsien and ku. The depth of the water, that is, ku, will be found by dividing this (remaining) rectangle (product) by 2; the length of the reed will then be obtained by adding 1 ch'ih to the depth of the water.

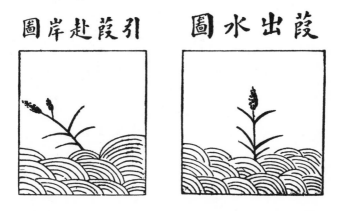

Figure 2.6a

Commentary: Employing the geometric-algebraic solution methods implied by the Han authors in their usage of the word rectangle to mean product and a modern algebraic symbolism for clarity, we consider the situation illustrated in Fig. 2.6b.

Figure 2.6b

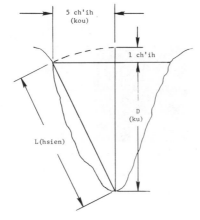

It is known that the sum of the areas of the squares erected on the legs of the right triangle should equal the area of the square erected on the hypotenuse (Fig. 2.6c):

$$(D + 1)^2 = D^2 + 25$$

Figure 2.6c

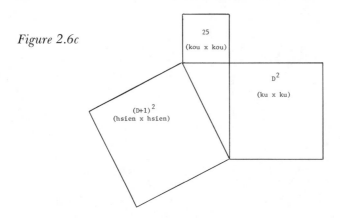

Manipulating the resulting squares to perform the indicated operations we obtain the configuration of Fig. 2.6d. The shaded portion of the square is equal in area to D^2, therefore the remaining two rectangles and the square must be equal in area to 25, i.e., $25 = 2D + 1$. If we subtract the square of 1 ch'ih as instructed, then the sum of the areas of the remaining two rectangles corresponds to 24, i.e., $2D = 24$. The area of one rectangle, D, will then correspond to the depth of the water, 12 ch'ih. To obtain the total length of the reed, we add the depth of the water, 12 ch'ih, to the length of the reed extending above the surface, 1 ch'ih, and obtain the desired result, 13 ch'ih.

Figure 2.6d

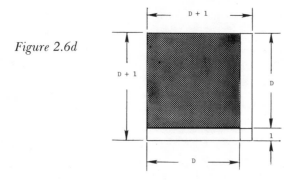

A more poetic rendering of this problem is found in the later writings of the Indian mathematician Bhāskara (1114–ca. 1182):

> In a certain lake, swarming with red geese, the tip of a bud of a lotus was seen a span (9 inches) above the surface of the water. Forced by the wind, it gradually advanced and was submerged at a distance of two cubits (approximately 40 inches). Compute quickly, mathematician, the depth of the pond.[8]

The similarity of these two problems even to the ratio of the two distances is striking. Such replication fuels a controversy concerning the influences of early Chinese mathematics on later Hindu writings.[9]

Figure 2.7a

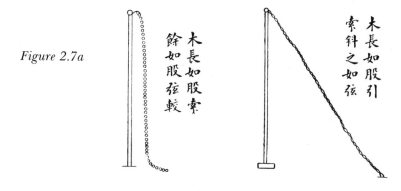

[7.] Given: A rope is tied to the top of a pole. The rope is 3 ch'ih longer than the pole. If we pull the rope (taut), the end will just touch the ground 8 ch'ih from the (base of the) pole. What is the length of the rope? (Fig. 2.7a)

Answer: 12⅙ ch'ih.

Method: Find the square of 8 ch'ih, divide the result by the difference of the rope and the pole, that is, 3 ch'ih. The length of the rope will equal one half the sum of the above quotient and 3 ch'ih.

Explanation: Let 8 ch'ih be the kou and the (desired) length of the rope be hsien, then the product of the sum and difference of hsien and ku is known to be the square of 8 ch'ih. Since the difference of hsien and ku is known to be 3 ch'ih, the sum of hsien and ku can be found to be the square of 8 divided by 3 or twice hsien minus 3 ch'ih equals the square of 8 divided by 3. Hsien is then found to be one half the sum of the square of 8 divided by 3 and 3.

Commentary: The conditions of the problem are represented in Fig. 2.7*b*. In the square shown in Fig. 2.7*d*, the shaded portion represents the product (hsien + ku)(hsien − ku) and must by the "Pythagorean" Theorem equal the square of kou, 64. Since (hsien − ku) = 3, then (hsien + ku) = 64/3. From Fig. 2.7*c* it can be seen that

$$\text{(hsien + ku)} = 2 \text{ ku} + 3 \qquad \text{and} \quad \text{(hsien + ku)} = 2 \text{ hsien} - 3$$

therefore

$$(2 \text{ hsien} - 3) = \frac{64}{3} \qquad \text{and} \quad \text{hsien} = \left(\frac{64}{3} + 3 \right) \div 2$$

Figure 2.7b

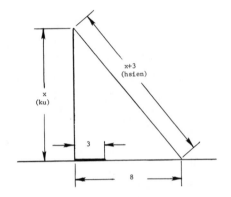

Figure 2.7c

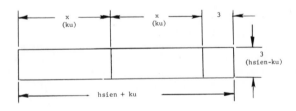

34

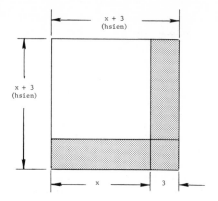

Figure 2.7d

[8.]   Given: The height of a wall is 10 ch'ih. A pole of unknown length
        leans against the wall so that its top is even with the top of the wall.
        If the bottom of the pole is moved 1 ch'ih further from the wall,
        the pole will fall to the ground. What is the length of the pole?
      Answer: 50.5 ch'ih.
      Method: Multiply the height of the wall by itself and divide this result
        by the distance the bottom of the pole is moved. The length of
        the pole is given by one half the sum of above result and 1 ch'ih.
      Explanation: Same as that for the rope problem.

Commentary:

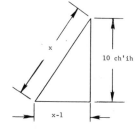

Figure 2.8

$$x^2 = 10^2 + (x - 1)^2$$
$$x^2 = 100 + x^2 - 2x + 1$$
$$x = \frac{100 + 1}{2}$$

Figure 2.9a[10]

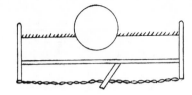

[9.]   Given: A wooden log is encased in a wall. If we cut part of the wood
          away, at a depth of 1 ts'un, the width of the (exposed) log measures
          1 ch'ih. What is the diameter of the log? (Fig. 2.9a)
       Answer: 2 ch'ih 6 ts'un.
       Method: Find the square of one half the obtained width, divide the
          result by the depth (of the cut) 1 ts'un. The diameter of the log
          will be the sum of the above result and 1 ts'un.
       Explanation: Consider one half the width, 5 ts'un, to be kou and one
          half the diameter to be hsien, then 1 ts'un should equal the dif-
          ference of ku and hsien. Proceeding as in problem [7], the radius
          is found to be the sum of half the width squared divided by 1 and 1.
          Doubling the radius we obtain the diameter.

Commentary: A variation of the preceding problem. Since x in
Fig. 2.9b equals the measure of the radii of the circle, $2x$ equals the
measure of the diameter, 2 ch'ih 6 ts'un.

Figure 2.9b

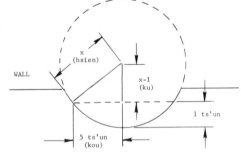

$$x^2 = (x - .1)^2 + (.5)^2$$
$$x^2 = x^2 - .2x + .01 + .25$$
$$.2x = .26$$
$$2x = 2.6$$

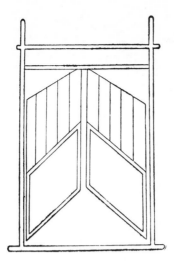

Figure 2.10a

[10.] Given: A two-door gate of unknown width is opened so that a 2 ts'un gap exists between the two doors. It is known that the (open doors') edges protrude 1 ch'ih from the door sill. What is the width of the gate?

Answer: 10 ch'ih 1 ts'un.

Method: The width of the gate will be equal to the sum of the square of 1 ch'ih and one half the 2 ts'un opening.

Explanation: Let 1 ts'un be kou, one half the width hsien, the difference of ku and hsien will be one half 2 ts'un. The width (of the gate) will be equal to twice hsien.

Commentary: Although the briefest of information is given concerning a solution procedure, it appears that the methods employed are similar to those used in the previous problems. From the information given it is known that hsien − ku = 0.2/2 = 0.1 and from Fig. 2.10*b* it is obvious that width = 2 hsien.

By constructing a physical model for the algebraic expression (Fig. 2.10*c*) (hsien − ku)(hsien + ku) = (kou)^2, using this equality and the fact that (hsien − ku)(hsien + ku) = 2 hsien we obtain

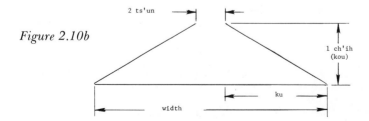

Figure 2.10b

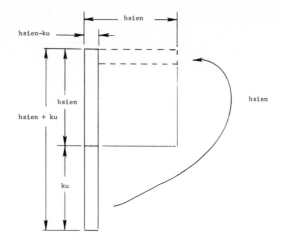

Figure 2.10c

$$2 \text{ hsien} = \frac{(\text{kou})^2}{(\text{hsien} - \text{ku})} + (\text{hsien} - \text{ku})$$

then

$$\text{width} = \frac{(1)^2}{0.1} + 0.1 = 10.1 \text{ ch'ih}$$

Figure 2.11a

[11.] Given: The height of a door is 6 ch'ih 8 ts'un larger than the width. The diagonal is 10 ch'ih. What are the dimensions of the door? (Fig. 2.11a)

Answer: Width = 2 ch'ih 8 ts'un, height = 9 ch'ih 6 ts'un.

Method: From the square of 10 ch'ih subtract twice the square of half of 6 ch-ih 8 ts'un. Halve this result and obtain its square root. The width of the door is equal to the difference of this root and half of 6 ch'ih 8 ts'un, and the height of the door is equal to the sum of the root and half of 6 ch'ih 8 ts'un.

Explanation: Consider the height (of the door) to be ku, width to be kou, and the diagonal hsien. The difference of kou and ku equals 6 ch'ih 8 ts'un. The square of ten is one hundred, double it and subtract the square of the difference of kou and ku. The square root will be the sum of the width and height, the width of the door will be half the difference of the square root and 6 ch'ih 8 ts'un. The height will be half the sum of the root and 6 ch'ih 8 ts'un.

Commentary: The conditions of the problem are illustrated in Fig. 2.11*b*. Three different geometric-algebraic solution schemes are suggested.

Figure 2.11b

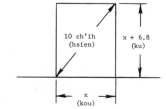

Figure 2.11c

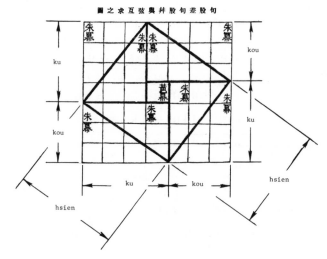

First, from Fig. 2.11c, the following equations can be derived:

$$(\text{kou} + \text{ku})^2 = 4(\text{kou} \times \text{ku}) + (\text{ku} - \text{kou})^2$$

$$(\text{hsien})^2 = 2(\text{kou} \times \text{ku}) + (\text{ku} - \text{kou})^2$$

$$(\text{kou} + \text{ku})^2 = (\text{hsien})^2 + 2(\text{kou} \times \text{ku})$$

$$(\text{kou} + \text{ku})^2 = 2(\text{hsien})^2 - (\text{ku} - \text{kou})^2$$

The last equation supplies us with a readily workable expression involving the unknowns and the given:

$$\text{kou} + \text{ku} = \sqrt{2(\text{hsien})^2 - (\text{ku} - \text{kou})^2}$$

$$\text{kou} + \text{ku} = \sqrt{2(10)^2 - (6.8)^2}$$

$$= \sqrt{153.76}$$

$$\text{width} = \frac{\sqrt{153.76} - 6.8}{2} = 2.8 \text{ ch'ih}$$

$$\text{height} = \frac{\sqrt{153.76} + 6.8}{2} = 9.6 \text{ ch'ih}$$

Figure 2.11c shows the *hsuan-thu* diagram of the *chou pei* and represents one of the earliest known proofs of the "Pythagorean" Theorem.[11] Admired for its simple elegance, it later found its way into the work of the Indian Bhāskara.

The Chinese ability to extract the square root of 153.76 attests to their highly developed computational proficiency. In this particular operation, the methods of the Han mathematicians surpassed those of other contemporary societies.

The second solution follows the geometric-algebraic methods of Chao Chün-Ching. From Fig. 2.11d it can be seen that the measure of the side of the square is equal to the length of 2 hsien. The area of the square is then equal to 4 hsien² and the following relations hold:

$$(2 \text{ hsien})^2 = 4 \text{ ku}^2 + 4 \text{ kou}^2$$

$$(2 \text{ hsien})^2 - 4 \text{ ku}^2 = 4 \text{ kou}^2$$

$$\sqrt{(2 \text{ hsien})^2 - 4 \text{ ku}^2} = 2 \text{ kou}$$

$$\frac{\sqrt{(2 \text{ hsien})^2 - 4 \text{ ku}^2}}{2} = \text{kou}$$

From the diagram it is also seen that

$$2 \text{ kou} = (\text{kou} + \text{hsien}) - (\text{hsien} - \text{kou})$$

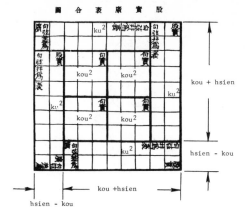

Figure 2.11d

While the geometric-algebraic identities are correct, their usefulness in this particular problem situation are limited.

Chao goes on to derive a similar expression for ku. From Fig. 2.11*e*, it can be seen that

$$(2 \text{ hsien})^2 = 4 \text{ ku}^2 + 4 \text{ kou}^2$$

$$(2 \text{ hsien})^2 - 4 \text{ kou}^2 = 4 \text{ ku}^2$$

$$\sqrt{(2 \text{ hsien})^2 - 4 \text{ kou}^2} = 2 \text{ ku}$$

$$2 \text{ ku} = (\text{ku} + \text{hsien}) - (\text{hsien} - \text{ku})$$

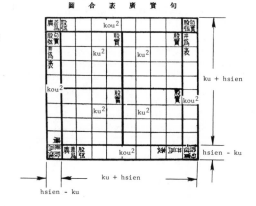

Figure 2.11e

Figure 2.12a

[12.] Given: A door and a measuring rod of unknown dimensions. Using the rod to measure the door, it is found that the rod is 4 ch'ih longer than the width of the door, 2 ch'ih longer than the height and the same length as the diagonal. What are the dimensions of the door? (Fig. 2.12a)

Answer: Width = 6 ch'ih, height = 8 ch'ih, length of diagonal = 10 ch'ih.

Method: Find the product of the difference of the rod's length and the door's width and the difference of the rod's length and the door's height. Double this product. The width of the door will equal the square root of this new product plus 2 ch'ih. The height of the door will equal the square root of the product plus 4 ch'ih.

Explanation: The height of the door is ku, the width kou, and the diagonal hsien. The difference of the hsien and ku is 2 ch'ih and the difference of hsien and kou is 4 ch'ih. The product of both their differences is equal to the products of hsien and hsien minus the products of ku and hsien and kou and hsien plus the product of ku and kou. Twice the product of the differences will equal the sum of the hsien squared, kou squared, ku squared, and twice the product of ku and kou minus twice the products of kou and hsien and ku and hsien, which is equal to the square of the sum of ku and kou minus hsien. Thus the square root of twice the products of differences will equal the sum of kou and ku minus hsien. If we add 2 ch'ih to this result, we obtain the width of the door. Similarly, if we add 4 ch'ih to this result, we obtain the height of the door.

Commentary: Using a modern notation, the problem situation is illustrated in Fig. 2.12b. The relations of the problem give rise to the equation

Figure 2.12b

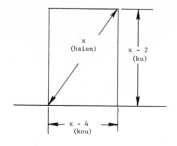

$$x^2 = (x - 2)^2 + (x - 4)^2$$
$$x^2 = 2x^2 - 12x + 20$$

or

$$x^2 - 12x + 20 = 0$$
$$(x - 2)(x - 10) = 0$$

thus

$$x = 2 \quad \text{or} \quad x = 10$$

Figure 2.12c

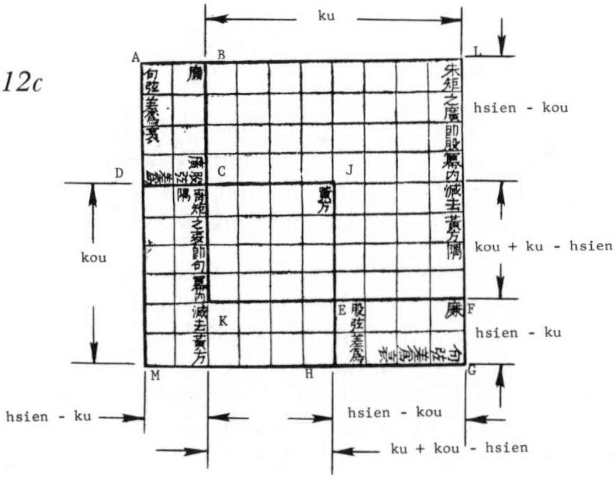

Practical considerations lead us to conclude that $x = 10$ is the solution we seek. The justification for the Chinese solution procedure is geometric-algebraic in nature. From Fig. 2.12c, the following equations can be derived:

43

area of rectangle ABCD = area of rectangle EFGH

$$= (\text{hsien} - \text{kou})(\text{hsien} - \text{ku})$$

(area of rectangle ABCD) + (area of rectangle EFGH)

$$= 2(\text{hsien} - \text{kou})(\text{hsien} - \text{ku}) = \text{area of square CJEK}$$

$$\overline{\text{CJ}} = \text{kou} + \text{ku} - \text{hsien}$$

$$\overline{\text{MD}} = (\text{kou} + \text{ku} - \text{hsien}) + (\text{hsien} - \text{ku}) = \text{kou}$$

$$\overline{\text{BL}} = (\text{kou} + \text{ku} - \text{hsien}) + (\text{hsien} - \text{kou}) = \text{ku}$$

Thus the lengths of kou and ku can be found.

Figure 2.13a

[13.] Given: A bamboo shoot 10 ch'ih tall has a break near the top. The configuration of the main shoot and its broken portion forms a triangle. The top touches the ground 3 ch'ih from the stem. What is the length of the stem left standing erect? (Fig. 2.13*a*).

Answer: $4\frac{11}{20}$ ch'ih.

Method: The square of 3 is 9. Divide 9 by the height of the bamboo shoot. The height of the plant left standing is found by subtracting the result obtained from 10 ch'ih and dividing by 2.

Explanation: Let 3 ch'ih be kou, the erect portion of the shoot ku, and the inclined portion hsien. The square of 3 ch'ih is equal to the difference of the squares of hsien and ku. The sum of hsien and ku is 10 ch'ih. The difference of hsien and ku is given by the quotient of 9 and 10.

Commentary: From Fig. 2.13*b*, we see

$$(\text{kou})^2 = (\text{hsien})^2 - (\text{ku})^2$$

$$= (\text{hsien} - \text{ku})(\text{hsien} + \text{ku})$$

$$= 9$$

44

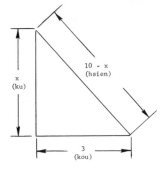

Figure 2.13b

We are given (hsien + ku) = 10, therefore

$$(\text{hsien} - \text{ku}) = \frac{9}{10} = 0.9$$

$$(\text{hsien} + \text{ku}) - (\text{hsien} - \text{ku}) = 9.1$$

$$2 \text{ ku} = 9.1$$

$$\text{ku} = 4\frac{11}{20} \text{ ch'ih}$$

This problem is also found in the ninth-century Sanskrit mathematical classic *Ganita-Sāra* (*Compendium of Calculation*) by Mahavira.[12] Still later in history it reappears in Florence in Philippi Calandri's "Arithmetic" (1491).[13]

[14.] Given: Two men starting from the same point begin walking in different directions. Their rates of travel are in the ratio 7:3. The slower man walks toward the east. His faster companion walks to the south 10 pu and then turns toward the northeast and proceeds until both men meet. How many pu did each man walk?

Answer: The fast traveler 24½ pu.

The slow traveler 10½ pu.

Method: The circuit traveled forms a right triangle. The three sides of the triangle are in the ratio given by the following magnitudes:

Northeast route, $\dfrac{(7^2 + 3^2)}{2}$

Southward route, $\dfrac{-(7^2 + 3^2)}{2} + 7^2$

East route, 3×7

The distance traveled to the northeast is then found to be

$$\left[10 \times \frac{(7^2 + 3^2)}{2} \right] \div \left[7^2 - \frac{(7^2 + 3^2)}{2} \right]$$

and to the east

$$10(3 \times 7) \div \left[7^2 - \frac{(7^2 + 3^2)}{2} \right]$$

Explanation: The faster rate (7) can represent the sum of kou and hsien and the slower rate (3) the ku. The sum of kou and hsien squared and ku squared all divided by 2 equals the sum of kou squared, twice the product of kou and hsien and hsien squared minus ku squared all divided by 2. This result simplifies to the product of hsien and the sum of kou and hsien, which is found to equal 29. The difference of the sum of kou and hsien squared and ku squared all divided by 2 and the sum of kou and hsien squared equals the sum of kou squared, twice the product of kou and hsien and hsien squared minus the sum of hsien squared and the product of kou and hsien. This result simplifies to the product of kou and the sum of kou and hsien, which is found to be 20. The product of ku and the sum of kou and hsien is found to be 21. These three results give the ratio of the three sides of the triangular circuit. The explanation is illustrated in Fig. 2.14*a*.

Figure 2.14a

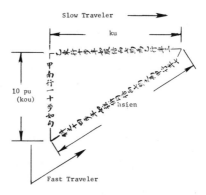

Commentary: The explanation of the solution technique is justified by the use of Fig. 2.14*b*.

$$\text{area of rectangle ABCD} = (kou + hsien)^2$$

It follows that

$$\text{area of rectangle EFGH} + \text{area of rectangle BHKJ} = (ku)^2$$

since the area of the square FGKM = $(hsien)^2$ and the area of the square EBJM = $(kou)^2$. The area of rectangle EFGH is seen to equal the area of rectangle JKLC, therefore it can be seen that

$$\text{area of rectangle AHLD} = (kou + hsien)^2 + ku^2$$

$$= (2 \ hsien)(kou + hsien)$$

46

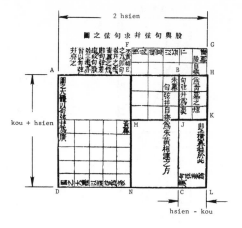

Figure 2.14b

Therefore

$$\tfrac{1}{2} \text{ area of rectangle AHLD} = \text{hsien(kou + hsien)}$$

which gives the term of the proportion relative to hsien. Similarly,

$$\text{(area of rectangle EHLN)} - \text{(area of rectangle BHLC)}$$
$$= \text{hsien(kou + hsien)} - (ku)^2$$
$$= (hsien)^2 + (kou)(hsien) - (ku)^2$$
$$= (kou)^2 + (kou)(hsien)$$
$$= \text{kou(kou + hsien)}$$

which supplies a term for the proportion relative to kou. The term for the proportion relative to ku is found to be ku (kou + hsien); thus

$$\frac{\text{hsien}}{29} = \frac{\text{kou}}{20} = \frac{\text{ku}}{21}$$

Replacing "kou" by the distance traveled southward, 10 pu, we obtain the distances traveled in the other required directions:

$$\frac{\text{hsien}}{29} = \frac{10}{20} = \frac{\text{ku}}{21}$$

$$\text{Northeast distance walked (hsien)} = \frac{29}{2} = 14\tfrac{1}{2} \text{ pu}$$

$$\text{Eastward distance traversed (ku)} = \frac{21}{2} = 10\tfrac{1}{2} \text{ pu}$$

$$\text{Total distance covered by faster man} = 10 + 14\tfrac{1}{2} = 24\tfrac{1}{2} \text{ pu}$$

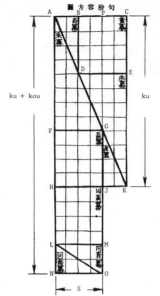

Figure 2.15a

ku + kou

ku

S

area of rectangle ACKH
= kou × ku = 72

area of triangle ABD
= area of triangle GKJ
= area of triangle LON
= area of triangle LOM

area of square BCED
= area of square FGJH
= area of square HJML

area of triangle AGF
= area of triangle DEK

[15.] Given: (A right triangle with) kou = 5 pu, ku = 12 pu
     What is the largest square which could be inscribed in the triangle?
   Answer: A square with side 3$\frac{9}{17}$ pu.
   Method: The side of the square will be given by the quotient of the
     product of 5 and 12 and the sum of 5 and 12.
   Explanation: See Fig. 2.15a. Use kou = 6 and ku = 12 to find a square
     whose side is an integer.

Commentary: The preceding explanation does not provide a proof
of the formula given but rather demonstrates that it works in a particular instance. The rectangle ACKH represents the product of kou
and ku. Its area is redistributed to form a rectangle ANOP representing the area (ku + kou) × S. Since the length (ku + kou) is known,
the desired length S for the side of the inscribed square can be found
by manipulation. It would appear that the formula was derived
using methods of proportion of which the Chinese were well aware.
     If we label the angle contained between the hsien and the kou of the
large triangle in Fig. 2.15b "θ," then the quotient of ku/kou given by the
authors is seen to be tan θ. Thus in their solution methods employing
proportions, Han mathematicians often made use of rudimentary
tangent functions. This solution method was so common at the time
that it bore a special name, *ch'ung-ch'a.*[z] The *ch'ung-ch'a* method of
solution will be used in several of the following problems.

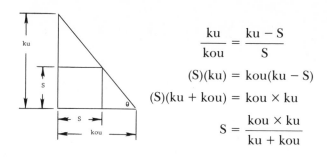

Figure 2.15b

$$\frac{ku}{kou} = \frac{ku - S}{S}$$

$$(S)(ku) = kou(ku - S)$$

$$(S)(ku + kou) = kou \times ku$$

$$S = \frac{kou \times ku}{ku + kou}$$

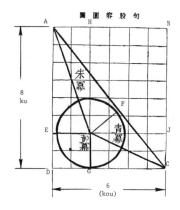

Figure 2.16a[14]

[16.] Given: (A right triangle with) kou 8 units long and ku 15 units. What is the largest circle that can be inscribed in this triangle? (Fig. 2.16a)

Answer: A circle with diameter of 6 units.

Method: Find the length of hsien from kou and ku. The diameter will be the quotient of twice the product of kou and ku and the sum of kou, ku, and hsien.

Explanation: The area of triangle ACD = ½(kou)(ku) which equals the sum of the areas of the following triangles AFO, AOE, FCO, CGO and the rectangle EOGD. Then

2(kou)(ku) = 4 area of triangle ACD

= 4(area of triangle AFO + area of triangle AOE)

+ 4(area of triangle FCO + area of triangle CGO)

+ 4(area of rectangle EODG)

In turn,

4(area of rectangle AHOE) + 4(area of rectangle OJCG)

+ 4(area of rectangle EODG) = 2(radius)(kou + ku + hsien)

49

therefore

$$2(\text{kou})(\text{ku}) = 2(\text{radius})(\text{kou} + \text{ku} + \text{hsien})$$

or

$$\text{diameter} = \frac{2(\text{kou})(\text{ku})}{\text{kou} + \text{ku} + \text{hsien}}$$

Commentary: Again the authors supply a demonstration of the validity of their solution formula. In the foregoing derivation the following relationships were used:

(1)   (ku × radius) = (area of AHOE) + (area of EOGD)

(2)   (kou × radius) = (area of EOGD) + (area of OJCG)

(3)   (hsien × radius) = (area of AHOE) + (area of OJCG)

Summing both sides of these equations and simplifying, we obtain:

$$\text{radius}(\text{ku} + \text{kou} + \text{hsien}) = 2(\text{area of AHOE})$$
$$+ 2(\text{area of OJCG})$$
$$+ 2(\text{area of EOGD})$$
$$= (\text{kou})(\text{ku})$$

therefore

$$\text{radius} = \frac{(\text{kou})(\text{ku})}{\text{ku} + \text{kou} + \text{hsien}}$$

$$\text{diameter} = \frac{2(\text{kou})(\text{ku})}{\text{ku} + \text{kou} + \text{hsien}}$$

In the derivation of (3) above, the authors indicate they are familiar with the congruence properties of right triangles.

Yang Hui[aa] in his thirteenth-century commentary (23) supplies a much simpler formula for finding the required diameter:

$$\text{diameter} = (\text{kou} + \text{ku}) - \text{hsien}$$

His "derivation" is interesting and worth examining.

If we form the rectangle whose height is ku and whose base is kou, we can arrive at the configuration seen in Fig. 2.16b. If this rectangle is now dissected and the pieces rearranged with an appropriate addition, the configuration of Fig. 2.16c results. The area contained within the rectangle of Fig. 2.16c is then

$$[\text{ku} + \text{kou} + \text{hsien}][(\text{kou} + \text{ku}) - \text{hsien}] = 2(\text{kou})(\text{ku})$$

50

Figure 2.16b

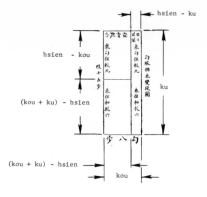

Figure 2.16c

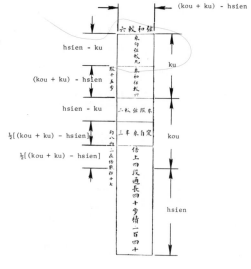

Since it has been established that

$$\text{diameter} = \frac{2(kou)(ku)}{kou + ku + hsien}$$

this identity can be used to simplify the formula for the diameter:

$$\text{diameter} = \frac{(ku + kou + hsien)(kou + ku) - hsien}{ku + kou + hsien}$$

thus

$$\text{diameter} = (kou + ku) - hsien$$

[17.] Given: A square walled city measures 200 pu on each side. Gates are located at the centers of all sides. If there is a tree 15 pu from the east gate, how far must a person travel out of the south gate to be able to see the tree?

Answer: 666⅔ pu

Method: This answer will be the quotient found by using the 15 pu as a denominator and half the width of the city squared as the numerator.

Explanation: The ratio of kou to ku is 15 to 100 (half side of city). Now the distance from the south gate to the east corner is a known kou (also half side of city). The problem remaining is to find the ku for this known kou, which we find to be 100 squared divided by 15.

Commentary: The formula appears to be derived by the use of simple proportions. Figure 2.17 illustrates the problem situation in question. Using the fact that triangles ABC and ECD are similar, the following proportion is employed to find the unknown distance, x:

$$\frac{15}{100} = \frac{100}{x}$$

$$x = \frac{(100)^2}{15}$$

$$x = 666\tfrac{2}{3} \text{ pu}$$

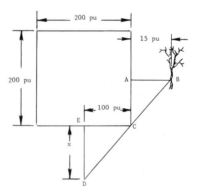

Figure 2.17

[18.] Given: A walled (rectangular) city measures 7 li in an east-west direction and 9 li in the north-south direction. There are gates at the centers of all sides. There is a tree 15 li from the east door. How many li must one walk out the south door before he can see the tree?

Answer: 315 paces or 1 li and 15 pu

Method: The answer will be given by the quotient where the numerator is equal to the product of the distance from the east gate to the southeast corner of the city and the distance from the south gate to the southeast corner, and the denominator is equal to the distance from the east gate to the tree.

Explanation: Same as the problem concerning the square city.

Commentary:

Figure 2.18

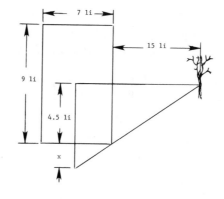

$$\frac{15}{3.5} = \frac{4.5}{x}$$

$$x = \frac{(4.5)(3.5)}{15}$$

Figure 2.19a[15]

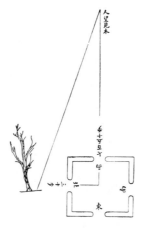

[19.] Given: A square walled city of unknown dimensions. There are gates at the center of each side. It is known that there is a tree 30 pu from the north gate, and standing 750 pu from the west gate one can see this tree. What are the dimensions of the city? (Fig. 2.19a)

Answer: 300 pu or 1 li

Method: Obtain the product of 750 and 30. The dimensions of the city will be equal to the square root of the product of 4, 750, and 30.

Explanation: As considered in the previous problem, the product of the two distances walked is equal to the square of half the measured length of the city wall.

Commentary: The answer is obtained by the use of simple proportion. Considering the situation depicted in Fig. 2.19b, we have

$$\frac{750}{x} = \frac{x}{30} \rightarrow x^2 = (30)(750)$$

$$2x = \sqrt{(4)(30)(750)}$$

Figure 2.19b

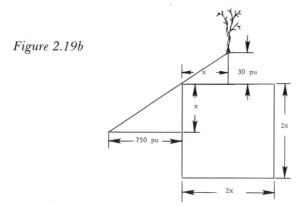

The fact that the authors double their result to obtain the desired solution by including a 4 in the radicand appears computationally inefficient and remains unexplained.

Figure 2.20a

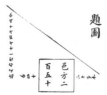

[20.] Given: A square walled city of unknown dimensions has four gates, one at the center of each side. A tree stands 20 pu from the north gate. One must walk 14 pu southward from the south gate and then turn west and walk 1775 pu before he can see the tree. What are the dimensions of the city? (Fig. 2.20*a*)

Answer: Each side is 250 pu long.

Method: Double the product of the distances walked north and west. This product should equal the product of the width of the city and the sum of the width of the city and the distances walked north and south. With this information the width of the city can be computed.

Explanation: Consider the distance traveled westward as the ku, from the tree to 14 pu south of the south gate the kou. The ratio of kou to ku is as 20 pu beyond the north gate is to half the width of the city; therefore the product of 20 and 1775 is equal to the product of half the width of the city and the sum of 20, 14, and the width of the city.

Commentary: Consider Fig. 2.20*b* with *x* representing the unknown length of the city wall. Triangles ABC and ADE are similar, lending themselves to the use of a simple proportion to develop a mathematical expression for *x*:

$$\frac{20}{\frac{1}{2}x} = \frac{20 + x + 14}{1775}$$

$$(20)(1775) = x(20 + x + 14)$$

$$x^2 + 34x - 71000 = 0$$

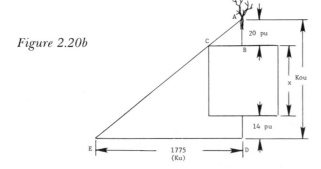

Figure 2.20*b*

The positive root of the quadratic equation gives the desired solution. Although the Chinese evidently employed a computational scheme called the *ts'ung fa*[ab] to extract the roots of this equation, the exact methods remain unknown to us.[16] Figure 2.20c, as supplied in Yang's commentary, illustrates the Chinese geometric-algebraic conception of the quadratic equation involved in this problem. Yang describes this figure as an aid to computation. Using the figure, we arrive at

$$x^2 + (20 + 14)x = 71000$$

but no further information on obtaining a solution is provided.

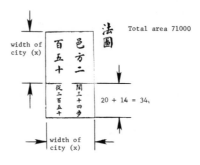

Figure 2.20c

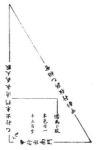

Figure 2.21a

[21.] Given: A square walled city measures 10 li on each side. At the center of each side is a gate. Two persons start walking from the center of the city. One walks out the south gate, the other out the east gate. The man walking south proceeds an unknown number of pu then turns northeast and continues until he meets the eastward traveler. The ratio of speeds for the southward and eastward travelers is 5 : 3. How many pu did each walk before they met? (Fig. 2.21a)

Answer: The man traveling south proceeds 800 pu from the gate, then continues 4887½ pu northeastward to meet the eastward traveler who has proceeded 4313½ pu from the east gate.

Method: Find the square of 5, find the square of 3, obtain their sum and halve it. This will give you the speed used along the northeast route. Subtract the northeast speed from the square of 5. The result gives the speed for the southward walk. Obtain the product of 3 and 5, which gives the eastward speed. Compute the product of half the width of the city (5 li) and the speed of the walk southward. Divide this result by the eastward speed and you will obtain the number of pu traveled southward (from the gate). The sum of half the width of the city and the above result will give the total distance traveled southward. To find the northeast distance obtain the product of the northeast speed and the southward distance; to find the distance traveled eastward, obtain the product of the eastward speed and the southward distance; divide each product by the southward speed, and you will obtain the distance walked northeast and east respectively.

Commentary: This problem is an extension of problem 14. The circuit traveled forms a right triangle whose sides are in the ratio

$$\frac{25 + 9}{2} \quad : \quad 25 - \frac{(25 + 9)}{2} \quad : \quad 3 \times 5$$

or

$$17 \quad : \quad 8 \quad : \quad 15$$

Figure 2.21b

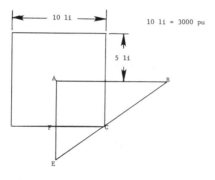

In Fig. 2.21b, we wish to find the distances $d(AB)$, $d(AE)$, and $d(EB)$. Using the fact that triangles BAE and CFE are similar, we can use the proportion

$$\frac{d(EB)}{17} = \frac{d(AB)}{15} = \frac{d(AE)}{8}$$

$$d(AE) = d(AF) + d(FE)$$

$$\frac{d(FE)}{8} = \frac{3000/2}{15}$$

$$d(\text{FE}) = \frac{(8)(1500)}{15} = 800 \text{ pu}$$

$$d(\text{AE}) = 1500 \text{ pu} + 800 \text{ pu} = 2300 \text{ pu}$$

then

$$\frac{d(\text{EB})}{17} = \frac{2300}{8}$$

$$d(\text{EB}) = 4887\frac{1}{2} \text{ pu}$$

and

$$\frac{d(\text{AB})}{15} = \frac{2300}{8}$$

$$d(\text{AB}) = 4312\frac{1}{2} \text{ pu}$$

Figure 2.22a

[22.] Given: A tree lies at an unknown distance from a point. Three additional points are marked on the ground. One point is colinear with the tree and the first point. The two remaining points lie to the right (of the line formed by the tree and two points). The distance between each point is 10 ch'ih. The tree is observed from the right rear point and the line of sight passes the forward right point at a distance of 3 ts'un. What is the distance to the tree? (Fig. 2.22a)

Answer: 333 ch'ih 3⅓ ts'un

Method: Find the square of 10 ch'ih and divide this result by 3 ts'un. The quotient will be 333 ch'ih 3⅓ ts'un.

Explanation: Consider 3 ts'un to be the proportionate kou and the distance between the two right points to be the proportionate ku, which is known to be 10 ts'un. Now the given kou is the distance between the right and left point. The problem is to find the new ku in respect to the given kou.

Commentary: A simple proportion is employed:

$$\frac{ku^*}{10} = \frac{10}{.3}$$

$$ku^* = \frac{(10)^2}{.3} = 333 \text{ ch'ih } 3\tfrac{1}{3} \text{ ts'un}$$

Figure 2.22b

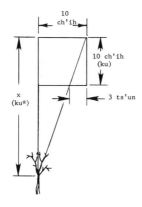

[23.] Given: A hill lies west of a tree whose height is 95 ch'ih. The distance between the hill and tree is known to be 53 li. A man 7 ch'ih tall stands 3 li east of the tree. If the tops of the hill and tree are aligned in the path of his vision, what is the height of the hill? (Fig. 2.23a)

Answer: 1649 ch'ih 6³³⁄₅₀ ts'un

Method: Obtain the height of the tree minus the height of the man. Multiply this difference by 56 ch'ih and divide by 3; the result plus 7 ch'ih gives the desired answer.

Explanation: The proportionate kou is equal to 95 ch'ih − 7 ch'ih or 88 ch'ih; the proportionate ku is equal to 3 li. The ku of 53 li plus 3 li or 56 li is known. Obtain the new kou with respect to the known ku. The height of the hill will then be the new kou plus 7 ch'ih.

Figure 2.23a

Commentary: Considering the situation illustrated in Fig. 2.23*b*, we have

$$\frac{x^*}{53} = \frac{95 - 7}{3}$$

$$3x^* = 53(95 - 7)$$

$$x^* = 1554.666$$

$$x = x^* + 95 = 1649.666 \text{ ch'ih or } 1649 \text{ ch'ih } 6^{33}\!/_{50} \text{ ts'un}$$

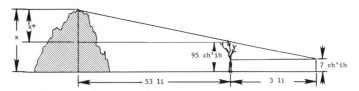

Figure 2.23b

[24.] Given: A deep well 5 ch'ih in diameter is of unknown depth (to the water level). If a 5 ch'ih post is erected at the edge of the well, the line of sight from the top of the post to the edge of the water (surface below) will pass through a point 4 ts'un from the opposite lip of the well. What is the depth of the well?

Answer: 57 ch'ih 5 ts'un

Method: Subtract 4 ts'un from the diameter of the well, 5 ch'ih; multiply the difference by the height of the post, 5 ch'ih. The depth of the well is equal to this product divided by 4 ts'un.

Figure 2.24a

Explanation: Consider the 4 ts'un to be the proportionate kou and the height of the post to be the proportionate ku. The given kou is equal to the difference of 5 ch'ih 4 ts'un and 4 ch'ih 6 ts'un. The dimension of ku with respect to the kou of 4 ch'ih 6 ts'un will be the depth of the well.

Commentary: From the situation as depicted in Fig. 2.24*b* we obtain

$$\frac{5}{x} = \frac{0.4}{(5 - 0.4)}$$

$$x = \frac{5(5 - 0.4)}{0.4}$$

$$x = 57.5 \text{ ch'ih or } 57 \text{ ch'ih } 5 \text{ ts'un}$$

Figure 2.24b

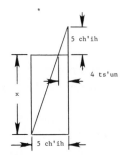

3

Conclusions

The Value of the *Kou-ku* as a Mathematical Thesis

The exposition of the *Kou-ku* does not comprise a deductive thesis on the right triangle and its mathematical properties, but it does display strong geometric insights and the existence of a high degree of computational proficiency in numerical solution techniques involving right triangles. A thorough understanding of the "Pythagorean" Theorem and properties of similar and congruent triangles is evidenced throughout the problem sequence. The authors exhibit a knowledge of geometry that, although not formalized, is quite explicit in problems such as number 16 where the essence of the solution method lies in the realization that the radii of a circle circumscribed by a triangle are perpendicular to the sides of the triangle at the points of tangency. The Chinese facility for root extraction, evidenced in several problems, appears to be superior to that possessed by contemporary societies. Also noteworthy is a knowledge of work with quadratics as demonstrated in problem 20 and the highly developed use of decimal systems of numeration and measure.

The organization of the *Kou-ku* in both scope and sequence possesses a certain pedagogical attractiveness. Definitions are given and formulas established. Problems progress in degrees of difficulty from a simple application of principles, to an analysis of imaginative geometric configurations, to even more complex exercises that require both situation analysis and a synthesis of previous solution techniques. These graduated degrees of abstraction reflect an appreciation of the human learning process on the part of the authors. Students of this text worked their computations assisted by a set of computing rods. This fact would seem to justify the mechanical nature of the solution schemes.

While by modern standards the *Kou-ku* format leaves much to be desired, it must be remembered that its form is typical of scientific tracts of its time. In fact, striking similarities exist between the picturesque problem situations of the *Kou-ku* and those of other ancient and even more modern texts. Babylonian tablets from the old period (1800–1600 B.C.) contain the following problem:

> A patu (beam) of length 0.30 (stands against a wall). The upper end has slipped down a distance 0.6. How far did the lower end move?[1]

Cultural and temporal transition to fourteenth-century Italy has changed the basic problem presentation very little:

> A spear 20 ft. long leans against a tower. If its end is moved out 12 ft., how far up the tower does the spear reach?[2]

Almost two hundred years later the first illustrated arithmetic book was published in Florence, with diagrams revealing several problem situations strongly reminiscent of the *Kou-ku*.[3]

Right triangle situations similar to those of the *Kou-ku* but with a different mathematical purpose are a recurrent theme in later Chinese mathematical literature. Li Chih,[ac] noted Sung algebraist, in his work, *Ts'-e-yuan hai-ching*[ad] (*Sea Mirror of the Circle Measurements*, ca. 1248), considers 170 problem situations whose solutions concern a circle inscribed in or circumscribing a right triangle (e.g., see Fig. 3.1).[4] Many of his exercises exhibit the influence of the *Kou-ku*'s problems:

> A person leaves the western gate (of a circular walled city) and walks south for 480 pu. A second person leaves the eastern gate and walks straight ahead a distance of 16 pu when he just begins to see the other person. What is the diameter of the city wall?

式 圖 城 圓

Figure 3.1

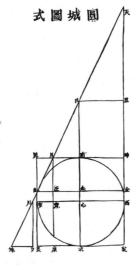

This problem situation gives rise to a fourth-degree equation, which can be readily reduced to a third-degree equation and solved. Ch'in

Chiu-shao,[ag] a contemporary of Li, used a similar problem to devise a tenth-degree equation![5]

> There is a round town of which we do not know the circumference or the diameter. There are four gates (in the wall). Three li from the northern (gate) is a high tree. When we go outside the southern gate and turn east, we must walk 9 li before we see the tree. Find the circumference and the diameter of the town. (See Fig. 3.2.)

Figure 3.2

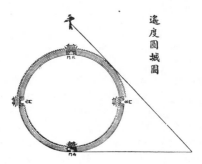

Allowing the diameter of the town to be x^2, the distance of the tree from the northern gate a, and the distance walked eastward b, we obtain the following equation:

$$x^{10} + 5ax^8 + 8a^2x^6 - 4a(b^2 - a^2)x^4 - 16a^2b^2x^2 - 16a^3b^2 = 0$$

Substituting the appropriate values for a and b, we have

$$x^{10} + 15x^8 + 72x^6 - 864x^4 - 11{,}664x^2 - 34{,}992 = 0$$

The diameter of the town is found to be 9 li. Thus the *Kou-ku* was to pervade Chinese mathematical thinking for centuries to come.

With this perspective established, the full worth of the *Kou-ku* of the *Chiu chang suan shu* can better be appreciated. Designed to satisfy the utilitarian surveying needs of early Chinese society, the *Kou-ku*'s problems and their implications warrant historic recognition as a classic in existing mathematical literature.

Right Triangle Solution Techniques of Ancient China

The particular solution rules used in the *Kou-ku* appear to have been obtained by two techniques, *chi-chu*[ah] and *ch'ung-ch'a*.

Chi-chu, the "piling up of squares," is used to obtain or substantiate the solution rules for eleven of the exercises considered. This method rests in a kind of geometric algebra, concretely based and appealing to intuition and readily adopted for modern classroom use with the aid of large-grid graph paper or geoboards. Such an approach to problem solving is also contained in early Greek works; however, the Greek methodology appears more abstract than its Chinese counterpart. The traditional Chinese proof of the "Pythagorean" Theorem, the *hsuan-thu* employed in obtaining a solution to problem 11, is admired by Coolidge in his history of geometry for its simple elegance. The high degree of facility with which the Han mathematician could employ such geometrically conceived algebraic solution schemes attest to a keen sense of spatial perception. One cannot help but wonder concerning the influence of manipulative Tangram exercises on the development of this ability.[6] Recent research on Tangrams attributes their origin to the Chou dynasty, predating the appearance of Liu's *Chiu chang*, and implies certain psychological and educational designs in their conception.[7] While the conjecture of Tangram exercises being more than mere children's games is intriguing, its resolution awaits the results of more detailed research on the intent and origin of Chinese Tangrams.

Obtaining geometric solutions by means of the *ch'ung-ch'a* technique was a standard mathematical practice in Han China. Indeed, at times the title of the last chapter of the *Chiu chang* has been given as *ch'ung-ch'a*.[8] This method is employed either directly or indirectly in obtaining solutions for seven of the problems considered. It is thought that in Liu's initial commentary nine more problems were added to the original twenty-four of the *Kou-ku*. These problems involved *ch'ung-ch'a* computations in the measuring of heights and distances employing surveyors' poles and sighting bars. It was appropriate that they were separated from the *Chiu chang* and incorporated into a surveying manual. They survive today as a separate Chinese mathematical classic, the *Hai-tao suan-ching*[ai] (*Sea Island Mathematical Classic*), so named for the first problem of the collection, which concerns an island:[9]

> There is a sea island that is to be measured. Two rods that are 30 ch'ih high are erected at a distance of 1,000 pu from each other, so that the rear rod aligns with the first rod and the island. When a man walks 123 pu back from the nearer rod, the top of the island is just visible through the end of that rod, if he tries to see with his eye brought on the ground. The summit of the island's peak is also seen to align with the end of the rear rod, when seen bringing the end in contact with the ground from a point 127 pu to the rear of that rod. It is required to know the height and distance of this island. (See Fig. 3.3.)

Ch'ung-ch'a was translated by Mikami as "double application of proportion" and further clarified by Needham as "a kind of substitute for trigonometric functions."[10] It is obvious from examining the applications of this method as existing in the *Kou-ku* that early Chinese

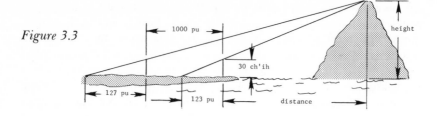

Figure 3.3

1000 pu

height

30 ch'ih

127 pu

123 pu

distance

mathematicians recognized the mathematical significance of the particular ratio formed by the sides of a right triangle opposite and adjacent to an angle of concern and could apply it in obtaining numerical solutions involving similar right triangles. Thus in their surveying needs, Han officials benefited from the utility of the tangent function. They did not, however, develop a theory of trigonometric functions, and it was not until the entry of Jesuits into China in the seventeenth century that the full power of trigonometric functions was understood.[11]

Was Pythagoras Chinese?

On the basis of the information examined in this study, what conclusions might be drawn concerning a sino-authorship of the "Pythagorean" Theorem?

The problems of the *Kou-ku* of the *Chiu chang* indicate that the Chinese had accumulated a wealth of experience in working with the right triangle in various mathematical situations well before the Christian era. No other ancient society, either among the predecessors or contemporaries of the Han people, could boast a similar level of accomplishment. The mathematical concerns of the ancient Chinese hydraulic society were utilitarian in nature and considered state priorities. In the China of this time, all mathematical and scientific knowledge was devised and recorded for the maintenance of the state. Scholars functioned for the court in a manner to be repeated in Renaissance Europe.[12] Under such conditions the universal validation of so basic a utilitarian mathematical relationship as the "Pythagorean" proposition would probably have warranted attention at a very early date. Since the Middle Kingdom had reached a comparatively high level of civilization a millennium before the emergence of a coherent Greek society, it would seem logical that serious Chinese considerations of the properties of the right triangle predated similar Greek efforts.

Finally, the *hsuan-thu* diagram of the *Chou pei* represents the oldest recorded proof of the "Pythagorean" Theorem. With simple elegance, the configuration and implications of this diagram appeal to both intuition and an aesthetic sense in proving that the sum of the squares of the sides of a right triangle equal the square of the hypotenuse. Its authorship, although anonymous, is indisputably Chinese.

If we now scrutinize the claim for Pythagorean authorship of the theorem that bears that name, we find that it rests mainly on tradition —a tradition that is founded on the realization that much of Western society's ethical, political, philosophical, and early scientific theories originated in the eastern Mediterranean region. But in actuality, little is known about Pythagoras the individual or his specific accomplishments. Accounts and facts are often clouded by mysticism and legend.[13] Smith, in his history of mathematics, states:

> According to our best evidence (1925) the familiar proposition that bears his name was known, as already stated in India, China and Egypt before his time, and all that can be claimed for him in relation to it is that he may have given the earliest demonstration of its truth.[14]

Neugebauer is not as kind, contending that "traditional stories of discoveries made by Thales or Pythagoras must be discarded as totally unhistorical." He then proceeds to substantiate his comment by discussing the relatively primitive state of Greek mathematics of the time—a condition which simply could not support the proposed discoveries.[15] Boyer in his recent history of mathematics prefers to consider contributions to the corpus of mathematical knowledge by Pythagoreans rather than Pythagoras, since he feels that many of the findings of the school were automatically attributed to its founder. The oldest available reference for consultation on the history of Greek mathematics is the *Eudemian Summary* (ca. A.D. 450) of Proclus. It assigns the authorship of the disputed theorem to Pythagoras. Considering the time span between Pythagoras' death and the publication of this book, one might wonder whether such as assertion also rests on tradition. Following the practices of the time. Pythagoras' teachings were oral; therefore, even if credence can be given to his claim of authorship, the exact derivation he may have used remains a matter of speculation. The proof popularly associated with him is the "windmill" or "bride's chair" proof given in Euclid's *Elements* (1.47), a proof that is intellectually taxing in comparison with the *hsuan-thu* diagram (see Fig. 3.4).[16] Thus it would seem that there exists ample room to challenge the accepted authorship of the first proof on the relationship of the sides of a right triangle.

While we hope a convincing case has been made for the primacy of right triangle theory in ancient China, further research will have to be conducted to resolve the basic question of this monograph. This research will have to be part of the much broader effort of trying

Figure 3.4

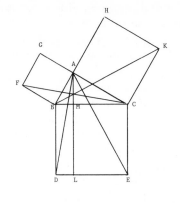

$$(AB)^2 = BM.BC$$
$$(AC)^2 = CM.BC$$
$$(AB)^2 + (AC)^2 = BC(BM + CM)$$
$$= (BC)^2$$

to understand the influence of early oriental mathematics and that of other cultures in general on the development of science and scientific thinking in the West. Francis Bacon duly noted the Chinese origins of the three inventions—gunpowder, printing, and the magnetic compass—that profoundly altered the history of Western civilization.[17] In a similar manner, when this research is completed, perhaps a future commentator might provide an equally revealing accounting of early China's mathematical accomplishments. Until that time, the question remains: Was Pythagoras Chinese?

Notes

Chapter 1

1. For a discussion of the instruments used to build the pyramids see Edwards, *The Pyramids of Egypt* (12) and Casson, *Ancient Egypt* (7).
2. An interesting discussion concerning the significance of the "Pythagorean Theorem" is given in Fredricks, *From Pythagoras to Einstein* (14).
3. Bronowski, *The Ascent of Man* (5), pp. 160–61.
4. Ibid., pp. 158–60; a variety of proofs have been suggested as Pythagoras' original. See, for example, Gittleman, *History of Mathematics* (16), p. 33; for the proof as given in Euclid's *Elements*, see Thomas, *Greek Mathematical Works* (39), 1: 179–85.
5. Heath, *The Thirteen Books of Euclid's Elements* (17), 1: 351–66; *A History of Greek Mathematics* (18), pp. 95–100.
6. The Chinese cosmological world view was dominated by the dualistic theory of the Yin and the Yang. Heaven and earth are opposites and are symbolized as such: Heaven is a positive force (Yang) represented by a hemisphere, earth is negative (Yin) represented by a square.
7. Translated by Alexander Wylie in *Chinese Researches* (43). Numbering of paragraphs by Wylie. Editorial clarifications in parentheses by Needham.
8. For an analysis of early China as an "hydraulic society" see Karl Wittfogel, *Oriental Despotism: A Comparative Study of Total Power* (New Haven: Yale University Press, 1957).
9. As given in Needham, *Science and Civilization in China* (31), 3: 25–27.
10. Ibid., p. 99.
11. Smith, *History of Mathematics* (36), 2: 483.
12. For a detailed discussion of the Celestial Element Method and its Greek counterpart, see Lam Lay Yong, "The Geometrical Basis of the Ancient Chinese Square-Root Method" (22) and Wang Ling and Joseph Needham, "Horner's Method in Chinese Mathematics" (42).
13. Coolidge, *A History of Geometrical Methods* (10), p. 21.
14. Smith, *History of Mathematics* (36), 2: 546.
15. A complete translation of this section is given by Lam Lay Yong, "Yang Hui's Commentary on the *Ying Nu* Chapter of the *Chiu Chang Suan Shu*" (23).
16. For more details on Chinese matrix problems see Struik, "On Ancient Chinese Mathematics (38) or Boyer, *A History of Mathematics* (3).
17. Kline, *Mathematics: A Cultural Approach* (21), p. 19.
18. Occasionally, Western writers have called attention to China's mathematical accomplishments. See, for example, Fellows, "China's Mathematics—the Oldest" (13) and Loria and McClenon, "The Debt of Mathematics to the Chinese People" (27) for two opposing views on this subject. Summarizing the difficulty in correctly assessing China's mathematical

accomplishment is an article by David E. Smith, "Unsettled Questions Concerning the Mathematics of China" (37).

Chapter 2

1. The *Yung lo ta tien* (*Great Encyclopedia of the Yung-lo Reign*) was compiled by order of the Emperor Chu Ti.[ac] Its contents include the corpus of Chinese knowledge up until the time of its writing. Over three thousand scholars labored at this task under the leadership of Hsieh Chin.[ad] The complete work contained 11,095 volumes of which only about 370 exist to the present day.
2. Problems are not numbered in the actual text.
3. See Van Der Waerden, *Science Awakening* (40), p. 76.
4. Translation as contained in Needham, *Science and Civilization in China* (31), 3: 22.
5. Ibid.
6. This theory is strongly advocated in Loomis, *The Pythagorean Proposition* (26), pp. 3–4; Loomis cites the Egyptian use of this triangle as a symbol of "universal nature." The base of length 4 represented Osiris; the remaining leg of length 3, Isis, and the hypotenuse represents Horus, the son of Osiris and Isis; Neugebauer cautions against the simplistic appeal of this theory in *The Exact Sciences in Antiquity* (33), p. 36.
7. *Pueraria hirsuta*, the oldest known Chinese fiber plant.
8. As translated in Coolidge, *A History of Geometrical Methods* (10), p. 17.
9. Needham discusses this matter in detail (31), 3: 146–50.
10. The diagram depicts a saw cutting into the log.
11. See Needham (31), 3: 95–96.
12. For a discussion of Mahavira's work see Smith, *History of Mathematics* (36), 1: 161–64.
13. Sanford, *The History and Significance of Certain Standard Problems in Algebra* (35), p. 77.
14. Chinese wood block print has an error in it—inscribed circle should have three tangents. Companion Figs. 2.16*b* and 2.16*c* are not drawn to scale.
15. This diagram employs the convention where the west is at the top of the figure and the north at the left. The usual Chinese directional convention was to indicate the cardinal points such that the south would be at the top, the north at the bottom, east to the left, and west to the right.
16. The *ts'ung fa* was a position on the counting board used for working the Celestial Element Method, therefore it appears that a variation of this technique was used for solving quadratic equations.

Chapter 3

1. Problem 9 of the old Babylonian text BM 85196 as translated in Van Der Waerden, *Science Awakening* (40), p. 76.
2. Cowley, "An Italian Mathematical Manuscript" (11).
3. As shown in Smith, *History of Mathematics* (36), 1: 255.

4. For a full discussion of the work of Li Chih see Ho Peng-yoke, "Li Chih" (20).

5. See Ho Peng-yoke, "Ch'in Chiu-shao" (19) and Libbrecht, *Chinese Mathematics in the Thirteenth Century* (25).

6. Tangrams, geometric puzzles consisting of assorted polygonal shapes that are to be joined together to form specific configurations.

7. Li and Morrill, *I Ching Games* (24); Gardner, "On the Fanciful History and the Creative Challenges of the Puzzle Game of Tangrams," (15). Gardner doubts any historical origins for tangrams.

8. Mikami, *The Development of Mathematics in China and Japan* (29), p. 34; Chang Yin-lin, "The Nine Chapters on the Mathematical Art and Mathematics in the Two Han Dynasties," (8).

9. See Sea Island problems contained in Medonick, *The Treasury of Mathematics* (28), pp. 462–64.

10. Mikami (8), p. 35; Needham, *Science and Civilization in China* (31), 3: 109.

11. The Jesuit Matteo Ricci published the first modern trigonometry book in China around 1607.

12. State patronage brought with it both benefits and disadvantages—the humanistic constraints of the Confucian ethic helped to stifle expressions of scientific creativity.

13. See Burkert, *Lore and Science in Ancient Pythagoreanism* (6).

14. Smith (36), 1: 72. Smith implies a strong oriental influence in both the theories and philosophy of Pythagoras.

15. Neugebauer, *The Exact Sciences in Antiquity* (33), p. 148.

16. For illustrations of this proof as reproduced in various cultures and time periods see Bergamini, *Mathematics* (2), p. 78. The philosopher Arthur Schopenhauer has termed this version of the proof "a mouse trap proof" and "a proof walking on stilts, nay, a mean, underhanded proof." Quoted by Morris Kline in *Mathematics: A Cultural Approach* (21), p. 50.

17. The history of science and technology in China is the subject of some recent books including Nakayama and Sivin, *Chinese Science: Explorations of an Ancient Tradition* (30), Needham, *The Grand Titration* (32), and Breuer, *Columbus Was Chinese: Discoveries and Inventions of the Far East* (4).

References

1. Berezkina, E.I. "Drevnekitajsky Traktat *Matematika v devjati Knigach*" (The Ancient Chinese Work Nine Chapters on the Mathematical Art). *Istoriko-matematiceskie issledovaniya* (1957), 10: 423–584.
2. Bergamini, David. *Mathematics*. New York: Time-Life Books, 1963.
3. Boyer, Carl B. *A History of Mathematics*. New York: Wiley, 1968.
4. Breuer, Hans. *Columbus Was Chinese: Discoveries and Inventions of the Far East*. New York: Herder and Herder, 1972.
5. Bronowski, Jacob. *The Ascent of Man*. Boston: Little, Brown, 1973.
6. Burkert, Walter. *Lore and Science in Ancient Pythagoreanism*. (Cambridge, Mass.: Harvard University Press, 1972.
7. Casson, Lionel, *Ancient Egypt*. New York: Time Incorporated, 1965.
8. Chang Yin-lin. "The Nine Chapters on the Mathematical Art and Mathematics in the Two Han Dynasties," *Yenching Hsueh Pao (Yenching University Journal of Chinese Studies)* (1927), 2: 301–12. (In Chinese)
9. *Chiu chang suan shu*. Taiwan: Commercial Press, 1965.
10. Coolidge, Julian Lowell. *A History of Geometrical Methods*. New York: Dover, 1963.
11. Cowley, Elizabeth B. "An Italian Mathematical Manuscript," *Vassar Medieval Studies*, 1923, pp. 379–405.
12. Edwards, I.E.S. *The Pyramids of Egypt*. Harmondsworth, England: Penguin, 1947.
13. Fellows, Albion. "China's Mathematics—the Oldest," *China Review* (1921), 1: 214–15.
14. Fredricks, K.O. *From Pythagoras to Einstein*. New York: Random House, 1965.
15. Gardner, Martin. "On the Fanciful History and the Creative Challenges of the Puzzle Game of Tangrams," *Scientific American* (August 1972), 231: 98–103A.
16. Gittleman, Arthur. *History of Mathematics*. Columbus, Ohio: Charles E. Merrill, 1975.
17. Heath, Sir Thomas L. *The Thirteen Books of Euclid's Elements*. 3 vols. Cambridge: The University Press, 1926.
18. ———. *A History of Greek Mathematics*. 2 vols. Oxford: Oxford University Press, 1921.
19. Ho Peng-yoke. "Ch'in Chiu-shao," *Dictionary of Scientific Biography*, vol. 3. New York: Scribner, 1973, pp. 249–56.
20. ———. "Li Chih," *Dictionary of Scientific Biography*, vol. VIII, pp. 313–20.
21. Kline, Morris. *Mathematics: A Cultural Approach*. Reading, Mass.: Addison-Wesley, 1962.
22. Lam Lay Yong. "The Geometrical Basis of the Ancient Chinese Square-Root Method," *Isis*, Fall 1970, pp. 92–101.
23. ———. "Yang Hui's Commentary on the *Ying Nu* of the *Chiu Chang Suan Shu*," *Historia Mathematica* (1974), 1: 47–64.
24. Li, H.Y. and Morrill, Sibley S. *I Ching Games*. San Francisco: Cadleon Press, 1971.

25. Libbrecht, Ulrich. *Chinese Mathematics in the Thirteenth Century: The Shu-shu chiu-chang of Ch'in Chiu-shao*. Cambridge, Mass.: MIT Press, 1973.
26. Loomis, Elisha Scott. *The Pythagorean Proposition*. Washington, D.C.: National Council of Teachers of Mathematics, 1968.
27. Loria, Gino and McClenon, R.B. "The Debt of Mathematics to the Chinese People," *Scientific Monthly* (1921), 12: 517–21.
28. Medonick, Henrietta O., ed. *The Treasury of Mathematics*. New York: Philosophical Library, 1965.
29. Mikami, Yoshio. *The Development of Mathematics in China and Japan*. New York: Chelsea, reprint of 1913 edition.
30. Nakayama, Shigeru and Sivin, Nathan, eds. *Chinese Science: Explorations of an Ancient Tradition*. Cambridge, Mass.: MIT Press, 1973.
31. Needham, Joseph. *Science and Civilization in China*. 7 vols. Cambridge: Cambridge University Press, 1954–.
32. ———. *The Grand Titration: Science and Society in East and West*. Toronto: University of Toronto Press, 1969.
33. Neugebauer, Otto. *The Exact Sciences in Antiquity*. New York: Dover, 1969.
34. Price, D.J. de Solla. "The Babylonian 'Pythagorean Triangle' Tablet," *Centaurus* (1964), 10: 219–31.
35. Sanford, Vera. *The Historical Significance of Certain Standard Problems in Algebra*. New York: Teachers College Press, 1927.
36. Smith, David E. *History of Mathematics*. 2 vols. New York: Dover, 1958.
37. ———. "Unsettled Questions Concerning the Mathematics of China," *Scientific Monthly* (1931), 33: 244–50.
38. Struik, D.J. "On Ancient Chinese Mathematics," *Mathematics Teacher* (1963), 56: 424–32.
39. Thomas, Ivor, trans. *Greek Mathematical Works*. 2 vols. Cambridge, Mass.: Harvard University Press, 1939.
40. Van Der Waerden, B.L. *Science Awakening*. New York: Oxford University Press, 1961.
41. Vogel, Kurt. *Chiu Chang Suan Shu: Neun Bucher Arithmetischer Technik*. Braunschweig, Germany: Friedrick Vieweg und Sohn, 1968.
42. Wang Ling and Needham, Joseph. "Horner's Method in Chinese Mathematics: Its Origins in the Root Extraction Procedures of the Han Dynasty," *T'oung Pao* (1955), 43: 345–88.
43. Wylie, Alexander. *Chinese Researches*. Shanghai: American Presbyterian Press, 1897.
44. ———. *Notes on Chinese Literature*. Shanghai: American Presbyterian Press, 1901.
45. Yang Hui. *Hsiang chieh chiu chang suan fa.*[aj] 2 vols. Shanghai: Commercial Press, 1936.

Glossary

a	周髀算經	s	大衍求數
b	弦圖	t	鮑澣之
c	術	u	永樂大典
d	勾股	v	康熙
e	九章算術	w	李淳風
f	始皇帝	x	弦
g	張蒼	y	趙君卿
h	劉徽	z	重差
i	方田	aa	楊輝
j	栗米	ab	定法
k	衰分	ac	朱棣
l	少廣	ad	解縉
m	天	ae	李藉
n	商功	af	測圓海鏡
o	均輸	ag	秦九韶
p	盈不足	ah	積矩
q	方程	ai	海島算經
r	正負數	aj	詳解九章算法